DIRTY SCIENCE

How Unscientific Methods Are Blocking Our Cultural Advancement

BOB GEBELEIN

Published by Robert S. Gebelein
Durham, North Carolina

Contact:
Robert S. Gebelein
18 Hawthorne Drive
Durham, NC 27712
Email: dirtyscience2@gmail.com

Large portions of this book are copied from *The Mental Environment*, copyright © 2007 Robert S. Gebelein.

My letter to the presidents and chancellors of top-ranked American colleges and universities, published as a "Guest Editorial" in *Journal of Parapsychology* (Gebelein, 2013), is wholly incorporated into this book.

"Fair use" is claimed for all quotations that illustrate unscientific methods.

ISBN PAPER 978-0-9614611-4-0
ISBN EBOOK 978-0-9614611-5-7

Library of Congress Control Number: 2018955734

Cover design by nirkri.

In memory of

Karen Baumer

who taught me

"Approach them with love."

Table of Contents

Acknowledgements

Many thanks to Alice Ruckert, Sally Rhine Feather, Rhonda Karg, Dean Radin, Pamela St. John, and Larry Burk for reading and commenting on my various manuscripts.

I am especially thankful to John Palmer for his comments and professional editing. But please do not fault him if you find things wrong with this book, because I didn't incorporate all of his editorial suggestions.

I will always appreciate the teaching of Karen Baumer, "Approach them with love."

To Sarasvati Ishaya, many thanks for her efforts to promote me, and for constantly urging me to speak more highly of myself.

To Amber Wells, I appreciate very much her professional support of my limited knowledge of sociology.

To Tyler Stevens, many thanks for his marketing suggestions and for acquiring a good cover design, and to nirkri for designing the cover.

And to the many people at The Rhine Research Center and in the field of parapsychology at large who contributed to this book with information, suggestions, and support, many thanks to all of you.

DIRTY SCIENCE

CHAPTER 1

The Politics of Physicalism

There are more things in heaven and earth, Horatio,
Than are dreamt of in your philosophy.
— William Shakespeare, *Hamlet*, Act I, Scene V

This book needs to be written. This story needs to be told. It is a story of corruption and abuse of power, of scientific ridicule and academic prejudice, of how the people we have trusted to give us accurate knowledge have betrayed that trust and blocked large segments of knowledge from our view. When we wake up and recognize it, this corruption will be seen to be as serious as the corruption in the Catholic Church that precipitated the Protestant Reformation or the corruption in the French government that brought on the Revolution. This corruption has not compromised too much the accuracy of scientists working in their fields of expertise, but its effect on the culture at large has been profound and crippling. Science has brought us up out of the Dark Ages, and now the scientific establishment has brought our culture back down into something like another Dark Ages. This scientific establishment has become something like an authoritarian priesthood with a rigid belief in a purely physical universe, a belief that was not determined by scientific methods but is defended and enforced with primitive, unscientific methods of social domination and manipulation such as ridicule and ostracism. Heretics are no longer burned at the stake. They are simply ridiculed out of existence. It is almost as effective and doesn't violate any health laws.

These extreme social pressures dominate not only the scientific community but also the entire academic community. It may be extreme to compare our times with the Dark Ages, but the comparison is certainly accurate for the academic community,

1

where the domination of universally held fixed beliefs restricts scientific inquiry and destroys academic freedom.

We are already deep into something like another Dark Ages. Somebody should have warned us that this was happening. Somebody should have written this book a long time ago. But I don't see this book anywhere. It is as if everybody was functioning in some kind of trance state.

This book needs to be written by somebody with the very highest scientific or academic credentials. Unfortunately, all of those people are caught up in the problem. They have been indoctrinated into the belief system. They would be shunned and ostracized and lose their whole social existence if they were to write such a book as this.

So it is left to those of us with lesser credentials to write this book and solve the problem. I have taken it upon myself to write this book because I am aware of the problem and I am not controlled by the scientific establishment. But I can't fix the problem by myself. I need your support. Ninety percent of Americans believe in God or a spiritual reality, and ninety percent of Americans have had a psychic experience. Have you been ridiculed by members of the scientific establishment because of your psychic experiences or spiritual beliefs? Have you lost your job or status within the academic community because of your interest or belief in non-physical phenomena? Do you feel that the scientific establishment is an obstacle to our exploration of the spiritual? I am hoping you will give this project your active support.

You don't have to be an expert in science to recognize the unscientific methods. They are easy to identify. When we learn to spot these methods, the scientists who use them will lose their credibility. They won't be believed or be taken seriously any more. We don't have to make any scientists change. It's as easy as that.

First of all, I think I need to say plainly and clearly that I have great respect for science itself. Science has so greatly increased the scope and accuracy of our cultural knowledge that I have no words powerful enough to describe it. Because of its success in giving us accurate knowledge, science has rightfully earned the very highest status in our culture.

But the high status of science has created a problem. Science has tapped into a human psychological need for "authorities," people who know all the answers. Scientists have acquired an aura of omniscience, of infallibility, of authority. They have become an unnamed "Them," whose opinion is "Truth," such that anybody who disagrees with them can be dismissed as mentally incompetent. Their opinions are believed and taken seriously, whether those opinions were arrived at via the scientific method or not. It is the scientific method that has given us our accurate cultural knowledge, not just the opinions of scientists. But because of their status as "authorities," scientists get away with using unscientific methods such as ridicule, authoritarian pronouncements, and power politics to put down, dismiss, and block the study of subjects beyond their ken, especially the psychic and the spiritual:

A Ph.D. scientist doing research on communications with spirit entities is shunned by his colleagues, who point at their heads to indicate that he is crazy.

(If they were doing science, they would ask to see his evidence, instead of behaving like eighth graders.)

A graduate student is advised not to write a paper on parapsychology, because it could be career-ending.

(This is creepy, like something out of a Kafka novel. Unknown people could be ending this student's career, for unknown reasons. Whatever happened to academic freedom?)

The editor who published a peer-reviewed paper on "intelligent design" was ridiculed as a "Bible thumper."

("Scientific ridicule" has become a familiar expression. But ridicule is not scientific.)

In 1988, the National Academy of Sciences issued a report that concluded: "The committee finds no scientific justification from research conducted over a period of 130 years for the existence of parapsychological phenomena" (Druckman and Swets, 1988, page 22).

Actually the committee DID NOT LOOK AT the research of 130 years. The report contained only 5 references to the peer-reviewed *Journal of Parapsychology*, which dates back to 1937, and no references either to the peer-reviewed British journal, which dates back to 1882, or to the peer-reviewed European journal. It does not even mention the name "J.B. Rhine," let alone refute his findings and the people who replicated them throughout the world. The scope of the study clearly did not support its sweeping conclusion.

This shoddy and obviously biased study, led by two known zealots against parapsychology, is in huge contrast to the very impressive scientific credentials of the members and leaders of the National Academy of Sciences. Their conclusion is now forever recorded in the annals of folly.

But the NAME of the National Academy of Sciences has given it credibility. People haven't bothered to look at the study, which in turn didn't look at the scientific evidence. All they have needed to know is that the National Academy of Sciences said this. That gives it "authority" and makes it "Truth."

These are only a few examples of dirty science. You may have examples in your own life of scientists ridiculing you, questioning your sanity, or assuming that their mentality is so much greater than yours that they can say, "You didn't see what you saw."

I am not saying that science itself is dirty. But when scientists, in the name of "science," backed by the high status of science, their own high scientific credentials, and prestigious names such as Harvard and Princeton and the National Academy of Sciences, use these unscientific methods, the name of science has been dirtied. So, even though this is not science, I feel justified in calling it "dirty science."

What do I mean by "scientific establishment?" Is this some kind of conspiracy theory? No, it isn't. I have a precise definition.

Scientists studying physical phenomena with the physical senses — the physicists, chemists, biologists, astronomers, and so on — have established themselves. They have become the scientific establishment.

In fact, they have been so successful in studying the physical reality that they have asserted that the only way to do science is to explore the physical reality with the physical senses, and they have taken this one step further, to decree that the only reality is the physical reality: The spiritual does not exist, psychic abilities are impossible, and the mind is nothing but the physical brain, which must be studied with the physical senses. But, in fact, these other aspects of reality do exist, and there are other modes of perception besides the physical senses with which to perceive them and make scientific observations.

The idea that there is no reality beyond the physical has come to dominate not only the scientific establishment, but also the academic community as well, at least in the United States. I used to call it "the physical hypothesis," but it is much more than a hypothesis. It has become an axiom of scientific orthodoxy, accepted at our major educational institutions as such an absolute inviolable truth, that so dominates all other thought, including scientific evidence, that people who express an interest in such subjects as precognition, telepathy, clairvoyance, remote viewing, psychokinesis, energy medicine, spirit entities, the power of prayer, reincarnation, levitation, or intelligent design are automatically dismissed as mentally incompetent, shunned by their colleagues, and denied publication, funding, and employment.

The foregoing may sound extreme, but I believe that it accurately represents the mental atmosphere that dominates our present-day academic environment. I invite academic people to present evidence showing my view to be flawed. The evidence as I see it shows that the subjects I have named in the preceding paragraph are rarely studied or taught at ranked colleges or universities in the United States.

Whenever I say this, people point out exceptions to the rule — a Master's thesis at Michigan, a discussion group at Harvard, a study of medical intuition at Duke, and a course in consciousness science at Duke. But these have been done quietly, without much publicity, and they haven't become part of the cultural awareness. You can't get a degree in parapsychology at any ranked college or university in the United States.

Further evidence that studies of such subjects are exceptions to the rule is that for the most highly publicized of such studies, people have felt the need to offer explanations as to why they were allowed:

Duke University was being newly created as an expansion of Trinity College in 1925-30, and thus was more open to J.B. Rhine and parapsychology than more established universities. The remote viewing at the Stanford Research Institute (SRI) was done at the request of the CIA and funded by government agencies.

Princeton Engineering Anomalies Research (PEAR) received its funding and building space year after year for research in "engineering anomalies" by a person with engineering credentials. The research on reincarnation begun by Ian Stevenson at the University of Virginia has been allowed to exist because it was funded by a grant specifically for religious and spiritual studies (from Chester Carlson, the founder of Xerox).

I am encouraged to have discovered Dr. Lisa Miller, who can be found by looking up "spiritual studies at Columbia." According to Wikipedia, "She founded the Spirituality Mind Body Institute at Teachers College, Columbia University, the first Ivy League graduate program in spirituality and psychology," apparently without any explanation of why this program was allowed to exist. Perhaps the barriers are softening already.

The assertion that there is no reality beyond the physical or what can be explained by known physical laws has been called "physicalism," "scientism," "reductionism," or "materialism." I prefer "physicalism," because it expresses most directly the idea of a belief in a purely physical reality.

For the assertion of a purely physical reality to be held with such absolute certainty, it must have been proved scientifically beyond any doubt. But I don't see that proof anywhere. Also, as scientists should know, "Nothing exists beyond the physical" is a logical proposition of the form, "There are no such things," which cannot be proved without a complete knowledge of everything that exists. It should be clear, with the discovery that "dark matter" occupies an estimated 96% of the physical universe, and the more recent revelation that dark matter doesn't really explain the observed phenomena, that we have nowhere near a complete knowledge of even the physical, much less the subtler aspects of reality that may exist.

And conversely, to disprove the proposition "There are no such things," it is necessary only to find one such thing. In 1958, I dreamed of my grandmother's death an hour before I received the telegram. That's all the evidence I ever needed in order to know there was a reality beyond what had been explained in my physics courses at Harvard. Billions of people worldwide have had experiences similar to mine.

Responsible researchers have been scientifically studying the psychic and the spiritual since the 1800s. The Society for Psychical Research in Great Britain was founded in 1882. J.B. Rhine in the United States began studying around 1930 what he called "extra-sensory perception" (ESP) with the most rigorous methods of physical science. I first heard about J.B. Rhine's work when I was at Harvard in the 1950s, although not in any of my courses. Parapsychology has been accepted as a science by the American Association for the Advancement of Science (AAAS) since 1969. The Alex Tanous Library at The Rhine Research Center has literally a "ton" of evidence for the existence of psychic and spiritual phenomena.

In recent years the legitimate findings of parapsychology have also been described in many books written for the general public, including *The Gift*, by Sally Rhine Feather, *Unbelievable*, by Stacy Horn, *Best Evidence*, by Michael Schmicker, *The End of Materialism*, by Charles Tart, *Let Magic Happen*, by Larry Burk, and *Supernormal*, by Dean Radin. These are only a few of the best.

Reincarnation researchers have developed a strict scientific protocol and have used it to uncover thousands of cases, some in the United States involving people who didn't even believe in reincarnation.

Research indicating that we are not limited to our physical bodies has been conducted by responsible people.

The power of prayer has been demonstrated in scientific experiments.

Energy healing has been demonstrated in scientific experiments.

I have a list of 144 peer-reviewed papers on intelligent design.

I have even seen evidence of levitation.

And yet there is still an extreme bias against the psychic and the spiritual by establishment scientists. They still assert, as first-principles of their belief system, that psychic abilities are impossible, that spirit entities do not exist, and that the mind is nothing but the physical brain.

Or perhaps I should use the scientific impersonal and say "It is assumed" within the scientific establishment that these things are so, and that these assumptions are considered basic criteria of rational thought. I don't want to pin the blame on anybody in particular, except for those few individuals I can actually quote. All scientists are innocent until proven guilty. But at the same time, all scientists must be part of the consensus, in order to make psychic and spiritual subjects laughable and not even debatable at ranked American colleges and universities.

Dirty Science is not repeating all the positive evidence. It is focusing instead on the illegitimate and unscientific arguments that are being used to block these legitimate findings. It is not seeking the parental approval of the scientific establishment. It is asking ordinary people such as yourself to recognize the unscientific arguments so that they no longer influence our culture.

Somehow, the idea that there is no reality beyond the physical has entered the minds of members of the scientific community and has captivated them, so that they insist that psychic abilities are impossible and spirit entities do not exist, even though these beliefs are not supported by scientific studies and even though there is a large and growing body of evidence of both psychic abilities and spirit entities.

This mental attitude, this belief in a purely physical universe, is a force in itself that needs to be removed if our culture is to advance. Researchers in the areas of the psychic and the spiritual have been seeking to win the approval of establishment scientists for their discoveries. But if establishment scientists refuse to look at the data, science can't be done. There is an extreme bias operating here. That bias must be removed before science can be done. That force blocking our cultural advancement is the subject of this book.

In social psychology we learn that when the legitimate arguments fail, people switch to illegitimate arguments. And so it is with physicalism: Since no legitimate scientific argument can possibly be made to support it, unscientific arguments have been used to defend and enforce it (again switching to the scientific impersonal so as not to put the blame on anybody except those I actually quote).

The first sign of an extreme bias is that establishment scientists simply refuse to look at the evidence. They just assume, since they believe that psychic abilities and spirit entities do not exist, that the evidence is flawed — that there were methodological errors, "dirty test tubes," or actual cheating. And where scientists refuse to look at the evidence, they aren't doing science.

Instead, these established scientists parade their credentials. They exploit their status as "authorities" by making authoritarian pronouncements such as "There are no such things" in reference to psychic or spiritual phenomena. And they are believed and taken seriously on the basis of their credentials, without having to have the necessary scientific evidence to back them up.

I put the word "authorities" in quotes, to mean "so-called 'authorities.'" This is a cultural phenomenon for which we have no word. We exist in an environment of "authorities" all our lives. When we are small children, our parents are as gods to us, all-powerful and all-knowing. Thus we are psychologically predisposed to believe that there are "authorities," people who know all the answers. This psychological conditioning is reinforced as we grow up, with teachers who know all the answers, while our job is simply to memorize what they teach, and preachers who claim to speak from the authority of God. In our adult lives we consult with experts — doctors, lawyers, plumbers, electricians, and so on — who know more about their respective fields than we do.

All these experts have credentials, from the Latin "credo," meaning "believe," to indicate that they can be believed. But credentials are only a heuristic, a rule of thumb, to indicate that these people have passed certain tests and are more likely to be accurate than people without credentials. It is no guarantee that they are accurate. It would take another expert to determine whether they really are accurate, and sometimes not even the experts know. Here is where people can and do set themselves up as "authorities," when they don't really know what they are talking about, and get people to believe them.

And so scientists, with their scientific credentials, can get people to believe them, with their authoritarian pronouncements, even though their statements have no scientific backing at all. This has been called "argument from authority." This more accurately should be called "argument from so-called 'authority.'" It is only the pretense of authority, and the people who do this are only pretending to be authorities. It would not be an argument from real scientific authority without the backing of replicated scientific studies. "Argument from authority" should actually be called "argument from status" or "argument from credentials."

The real criterion by which scientific statements should be judged is on their accuracy, and not on the basis of credentials. Actually, credentials create a bias, a judgment that the person with the greater credentials is more likely to be accurate. To eliminate this bias, peer reviews are sometimes conducted without knowing the author's name. Scientists know this, and some scientists deliberately use their credentials to create a bias in their favor.

Another unscientific method and the key to enforcing the belief in physicalism is argument by ridicule. There is ordinary "soft" ridicule, and then there is "hard" ridicule. Soft ridicule is only laughter. It is distracting, but only a nuisance, like dogs barking.

But since I have become involved in the field of parapsychology, I have discovered that there exists a phenomenon which I call "hard ridicule," where people are deprived of publication, funding, or employment because they are interested in the wrong subjects or hold the wrong ideas. Heretics are not threatened with loss of life, as they were in the Dark Ages, but with loss of livelihood. Also they lose their whole social existence as their colleagues shun them. This is equivalent to what excommunication was in the Dark Ages. It is as close to death as anybody but a government is allowed to do to a person in these present times.

And then we have power politics. In the Preface to my Delta edition of *Worlds in Collision* by Immanuel Velikovsky, the author explains that the book was originally published by Macmillan, a company that did a large proportion of its business selling textbooks, but that members of the scientific community boycotted Macmillan, refusing to buy their textbooks as long as they published *Worlds in Collision*. Therefore Macmillan was pressured to sell its rights to the book to Doubleday, a company that was not so dependent on textbooks. I suspect that the same kind of stuff is going on all the time, but it is not all as well publicized.

These are some of the unscientific methods — refusal to look at the evidence, authoritarian pronouncements (so-called "argument from authority"), ridicule (soft and hard), and power politics — used by the scientific establishment to defend and enforce the belief in physicalism and block the study of the psychic and the spiritual.

Because our Western civilization has limited its explorations to the physical, the generation who discovered the spiritual had to go to India to learn more about the spiritual reality. The New Age culture which they developed is now ridiculed and shunned by the Western academic establishment.

In my own personal experience, I didn't have to go to India. All my influences were from Western civilization. I discovered the spiritual and the key to a new civilization by going through Freudian and Jungian psychotherapy and dream analysis. I then learned more about the spiritual from Edgar Cayce and Eklal Kueshana. Freud, of course, was accepted for a long time by the academic community. But now Freud is "out." And all these other influences have been dismissed by the academic community. I was hoping that academic people would read and evaluate my ideas for a new civilization, but it seems that they were so blocked by prejudices against my major influences that they never got around to reading and evaluating my own ideas. Thus Western civilization has been deprived of my positive contribution without even being aware that it exists.

Because the United States has fallen behind other developed countries in math and science, President Obama in recent years tried to encourage young people to study these subjects. But the Hippies rejected science years ago, because of its adherence to a purely physical reality, and President Obama was trying to convince their grandchildren, who I'm sure were already aware of its limitations. Cleaning up the invalid mental attachments of the scientific community will not only remove barriers to our cultural progress, but it will also improve the quality of physical science itself, by attracting more of the best and brightest of our youth.

In more recent years, increasing numbers of people have been questioning the authoritarian pronouncements of scientists. This is reflected in Donald Trump's anti-science. These people are on the right track, but they don't quite have it right. Scientists doing science within their fields of expertise, such as global warming, are probably accurate. But there is no reason to believe scientists just because they are scientists and speak with "authority."

Scientists can get away with unscientific methods because they have the power. But we have given them this power by believing everything they say, and we can take it away, simply by questioning their authority.

Again, it is the scientific method that has given us our accurate cultural knowledge, not just the opinions of scientists. We need to distinguish between the methods of science, which have given us our accurate knowledge, and the methods of status, which people have used to dominate other people since before we even existed as human beings.

As a first approximation to knowing the difference between what is science and what is not science, undergraduate courses in critical thinking teach us the difference between "legitimate" and "illegitimate" arguments. And even if not all "legitimate" arguments qualify as scientific, we can be certain that all "illegitimate" arguments are unscientific. That distinction alone is enough to identify most of the unscientific arguments described in this book. For those who haven't taken the course, I am going to give a brief description of scientific and unscientific methods in the next chapter.

Any scientific opinion, to be truly scientific, must be based on scientific studies, not just one, but more than one, all reaching the same conclusion, meaning that they are "replicated." If the scientist does not cite replicated scientific studies, then his/her opinion is not scientific.

The third question to ask is, "Is the scientist qualified in this particular field?" Most of the negative opinions about parapsychology are by people totally unqualified in that field.

These three questions, then, will eliminate just about all the unscientific opinions:

1. Is this an illegitimate argument?
2. Is this opinion based on replicated scientific studies?
3. Is this person qualified in this particular field?

This takes a little more work than simply believing scientists because of their status as scientists, but it will go a long way towards untangling the huge snarl of false belief that now dominates Western civilization.

I think it is most important to recognize that scientists are not "authorities" who know all the answers about everything. A "scientist" is generally defined as a person with a Ph.D. degree in a scientific field. Scientists are knowledgeable in their particular fields of study, and earn their Ph.D. degrees by becoming expert in a very narrow segment of that field. They qualify as "expert" only in that very narrow segment, and outside of the broader field I would question whether their opinions should carry any weight at all. Because they are scientists, they are more qualified to say what is "science" than poets, priests, or rock stars. But that is all. Outside of those fields that they have studied, the opinions of scientists are largely uninformed.

For example, I have read about experiments by physicists trying to disprove astrology (CSICOP, 1996). Studies testing the accuracy of astrology should be conducted by people knowledgeable in astrology. I don't think that physicists have the necessary knowledge of astrology to perform accurate studies in that field. I am wondering whether physicists would accept experiments in physics performed by astrologers?

Similarly, the committee that evaluated parapsychology in the name of the National Academy of Sciences did not include a single member who was expert in parapsychology.

And the published opinions of eminent biologists Francis Crick and Edward O. Wilson on the nature of dreams are an embarrassing display of ignorance and arrogance.

Outside of the domain of the physical, in the areas of the psychic and the spiritual, physical scientists are totally unqualified. If they don't even believe that such things exist, that implies that they have absolutely no knowledge about them.

I use the expressions "physical science" and "physical scientists" to mean scientists studying physical phenomena with the physical senses. These scientists would say that these expressions are redundant, because this is the only way that science can be conducted. Even the *Encyclopedia Britannica* online now defines "science" as being conducted with the physical senses. This is a limited and limiting view, as I am going to explain.

How did scientists become locked into this belief system? It is not an easy question to answer. It involves psychological and sociological factors, and another factor called "how things happen."

I can understand why physical scientists might see only physical reality. Scientists studying physical phenomena with the physical senses are not likely to find anything but physical reality. So I suppose that after centuries and billions of observations, finding nothing but the physical, some of them might be led to believe that there is no reality beyond the physical. But remember that they limited themselves to the study of the physical in the first place.

It would be more accurate, then, to say that their view is LIMITED to the physical. But physical scientists have acquired such an aura of omniscience that it is some kind of blasphemy to say that their view is limited. But it was artificially limited in the first place. They didn't ask about ghosts or angels. They simply ignored these subjects and focused on physical things observable with the physical senses.

And if some people now want to ask about things such as ghosts and angels, and apply scientific methods to those questions, the members of the scientific establishment are totally unqualified to voice a scientific opinion on such subjects, having ignored them in the first place.

But again, to suggest that the scientific establishment is "not qualified" is some kind of blasphemy. The scientific establishment decides who is qualified. But I think that all establishment scientists would recognize that it is more accurate to say that their view is limited to the physical than to say that nothing exists beyond the physical.

The main support for physicalism has been Darwin's theory of the origin of species. His theory was quickly picked up and expanded to speculate on how life itself might have evolved by unintelligent design, without the need for a Creator. And it was quickly accepted as absolute truth supporting the idea of a purely physical universe, without being proved scientifically. But this theory is only speculation on how life MIGHT have evolved, and not scientific certainty of how life DID evolve. I am going to show in a later chapter how biologists have actually disproved Darwin's theory, but can't admit it because their whole belief system rests upon it. Also Darwin's theory as amended cannot be proved by biologists, because it carries with it the implied assertion that life DID NOT evolve by intelligent design, and it is beyond the scope of biology to demonstrate this. The dogma of unintelligent design, like physicalism itself, has not been proved scientifically, and so has to be defended and enforced by unscientific methods.

I can see why scientists in physical fields such as physics, chemistry, biology, and astronomy might at least pay lip-service to physicalism. It doesn't affect their scientific work in their fields of expertise, while opposing it could jeopardize their status as scientists. Yes, a dedicated physical scientist can treat the non-physical as if it doesn't exist, but to make the leap and assert that it really doesn't exist is really bad logic.

Actually, physical scientists studying physical phenomena with the physical senses do find things, fairly often, that don't conform to known physical laws. Instances of synchronicity are dismissed as "coincidence," although highly improbable. Unexplained healing is labeled "the placebo effect," as if they knew all about it and it was explained by physical science. Similarly, the origin of species is explained as "random" mutation, as if randomness was an explanation of the process causing the mutation. Regarding non-conforming observations, Albert Einstein is reported to have once said, "Even if I saw a ghost I wouldn't believe it" (Horn, 2009, page 33). With that kind of attitude, of course, the physical scientists can edit out anything that doesn't fit their belief system.

But there are other aspects of reality than the physical, and other senses with which to observe them. If we want to learn about these other aspects of reality, we might want to study them scientifically.

Our mental processes — our thoughts, emotions, memories, and dreams — are important influences in our human existence. They cannot be observed with the physical senses. They are observed with a different set of perceptions, which I call "the mental senses." The scientific study of the mind must start with the evidence of the mental senses, with whatever difficulties that this presents.

The early psychologists did not recognize the mental senses as such. They were working with something they called "introspection," which has a convoluted definition, combining perception, thought, and even opinion. They were not very successful. If they had been able to focus purely on the perceptions of mental events, they would have done better.

Freud and Jung and the early psychoanalysts, on the other hand, were very successful, building up a huge body of accurate knowledge based upon the evidence of the mental senses (although they did not recognize these senses per se). But this body of knowledge and the scientific study of the mind have been swept away by the dogmatic insistence on physical evidence by the scientific establishment.

With psychic and spiritual phenomena, instead of simply asserting that these things don't exist, it would be more scientific to see if there is any evidence of their existence. But the evidence of these things, obtained by qualified and responsible scientists in the established way with the physical senses, is swept away by the scientific establishment, simply because They are established and the others are not.

The scientific study of non-physical subjects is relatively new, compared to the study of physical subjects, especially physics and medicine, which date back to ancient Greece in Western civilization. Psychoanalysis 100 years after Freud has had only as much time to develop as medicine 100 years after Hippocrates. The non-physical scientists are not established. They are not part of the scientific establishment. They are treated with prejudice by the scientific establishment, very much as new immigrants to America have been treated by the social establishment, as in "Irish need not apply." But that is status-snobbery. It has nothing to do with scientific thinking or accuracy.

Recent surveys have shown that the vast majority of people in our Western culture have had psychic experiences, and that the percentage of people who have reported such experiences has been increasing in the 21st century. I am grateful to Carlos Alvarado for bringing many of these studies together in one concise report:

In a telephone survey conducted by the Institut fur Grenzgebiete der Psychologie und Psychohygiene (IGPP) in Germany in 2000, more than half of the people gave an account of personal exceptional experiences.

In 2006, Dr. Erlendur Haraldsson conducted a survey of people in Iceland, where 70% of the men and 81% of the women reported some psychic experience. These percentages had increased from 1974, when 59% of the men and 71% of the women reported psychic experiences.

In Switzerland in 2014, in a sample of 1580 people selected carefully to be representative of the Swiss general population, 91% of all participants had experienced at least one "exceptional experience" (EE), such as precognition, supernatural appearances, or déjà vu (Alvarado, 2014).

So you can't really call these experiences "paranormal." They are normal. Somebody has already said that. Cris Putnam, in promoting his book, *The Supernatural Worldview*, said "Paranormal is the New Normal." He reported an online survey of more than 4000 Americans between 2006 and 2011, with 90% reporting at least one "paranormal experience" (Putnam, 2014).

And yet an entrenched scientific establishment denies that this is possible, in the face of all evidence.

According to Gallup polls over the years, roughly 90% of Americans believe in God or a higher power (Gallup, 2016).

In contrast to this, a survey of elite scientists (members of the National Academy of Sciences) conducted in 1996 by Edward J. Larson and Larry Witham showed that roughly 70% of these scientists DID NOT believe in God or human immortality (Larson and Witham, 1998).

There is an extreme bias operating here.

A survey conducted in June 2009 by the Pew Research Center finds less of a bias, but still a definite bias. (Perhaps this is because the scientists they selected, members of the American Association for the Advancement of Science (AAAS), were not all as "elite" as the members of the National Academy of Sciences selected for the 1996 survey.) This poll showed that 51% of the scientists said that they believed in God, a universal spirit, or a higher power, in contrast to 95% of Americans who believed in some kind of deity or higher power, according to a survey done by the Pew Research Center in 2006 (Pew Research Center, 2009). However you look at it, there is a very large bias that separates the beliefs of scientists from those of the general public.

Who has the bias? Without even asking the question, it is assumed by members of the scientific establishment that the general public has the bias. Persons of status can get away with this.

They can project their faults on persons of lesser status, because those persons have no power to contradict them. But isn't it more likely that members of the scientific establishment have the bias, with the extreme pressures that are on them to conform to a belief in a purely physical reality? This possibility is simply by-passed by the scientific establishment. Because They are scientists, and therefore assumed to be unbiased, it is simply assumed that the general public has the bias. The explanation given for this is that those of us who are not scientists are uninformed, unintelligent, superstitious, and/or delusional (of course without the necessary psychological testing).

I am sure that most scientists, if they thought about it, would recognize this as gross class prejudice. And likewise, if they thought about it, I am sure that scientists would recognize the bad logic in the assertion, "There is no reality beyond the physical." And again, if they thought about it, they would recognize the unscientific nature of authoritarian pronouncements, refusal to look at the data, hard ridicule, and power politics. But they don't think about it.

What is going on here? This is not scientific thinking. This is more like in-group thinking.

"Science" is a method of acquiring knowledge through observation and proof. "Science" is also a social group, of people who practice the scientific method. And not all of the opinions, attitudes, and beliefs of people in this social group are arrived at by the scientific method. Some of their views are simply conformity to the norms of their social group.

If you belong to a social group, any group, you must conform to its norms and at least pay lip service to its opinions, attitudes, and beliefs. If you don't, you will first be gently instructed, then you will lose status in the group, and then you will be expelled from the group.

The persons with the highest status in the group are allowed some degree of nonconformity, to innovate. The persons with the lowest status in the group are also allowed a degree of nonconformity, but they are ridiculed as fools and are the butt of jokes, thus serving a useful social purpose.

The social group is known in sociology and social psychology as the "in-group." Everybody outside of the group is known as the "out-group." Obviously there are many other social groups, so I have seen the plural "out-groups." But to the members of the in-group, all persons outside their group are viewed equally as "not-in," and therefore equally as one out-group. The status of all those in the out-group is lower than the lowest in the in-group, and therefore, even more than the lowest in the in-group, they can be the target of jokes and objects of ridicule.

In-group thinking is viewed as perfect, and out-group thinking, where it is different, is viewed as flawed. Members of the in-group are seen as superior to people in the out-group, who have faults that in-group people don't have.

I am grateful to sociologist Amber Wells, Ph.D., for verifying this description of in-group thinking and providing the following professional definition and clarification:

> Sociologists and social psychologists use the term in-group to describe a social group whose membership depends on a shared identity; this can range from a preference for a specific diet or book genre or even support for a political party. Members of an in-group can become so homogeneous in their relationship to this identity that even a slight deviation from it could result in their rejection/removal from the group and elicit the derision of its members. In-group thinking, therefore, can be especially vulnerable to confirmation bias and other logical fallacies. Outgroup members are those who do not conform to the collective identity around which an in-group is organized. Their opinions/voices/perspectives are often considered to be less valuable and less worthy of consideration than those of in-group members.

I have seen these attributes of in-group behavior demonstrated with small groups, and I have extrapolated those findings to include the very largest social groups, such as the Catholic Church, the Hippies, the scientific establishment, and the whole academic community. Certainly we are aware of the in-group thinking of large groups in cases of religious prejudice, war, and genocide. But somehow in my Harvard education the idea was implanted in my head that academic people were free of prejudice, and scientific people were so completely unbiased that the word "objective" was used to describe their thinking.

So I am asking scientific and academic people to examine the opinions, attitudes, and beliefs of their own social group. Do they fit the description of in-group thinking that I have given? In order to answer this question, a degree of self-knowledge is necessary: Why do you conform to beliefs of your social group that are not based on scientific studies or good scholarship?

And I think that the answer to that is that peer-group pressure is more powerful than peer review. The normative pressures exerted upon people to conform to in-group thinking are more powerful than the intellectual forces that produce accurate science or scholarship. If one's scientific theory is wrong, it is debated and rejected. But if one does not conform to the in-group thinking, one's entire professional and social existence can be rejected. An important part of the education of scientific and academic people is indoctrination into in-group norms, although this is not so much part of the course material, but is done more with social signals such as smiles and frowns.

Scientists operate with the greatest possible accuracy within their own fields of expertise. It is mostly in dealing with areas of knowledge outside their own, as physical scientists talking about non-physical subjects, that they use the unscientific arguments of physicalism. This is the standard pattern of in-group thinking, with its prejudices against the out-group.

For example, the Catholic Church has recorded thousands of miracles, and the Christian Science Church has recorded thousands of miraculous cures. But all this evidence is dismissed as "anecdotal evidence," because it wasn't acquired by people in the scientific in-group, who are automatically assumed to be accurate, whereas people in the out-group are not. Also, when scientists insist that observations must be made by "trained observers," they exclude police officers, radar operators, and air traffic controllers, who are trained observers but not members of the scientific in-group. All this is already gross class prejudice, without going so far as to call anybody "delusional."

It all depends on the subject matter. Procedures that would be recognized as good science if applied to physical subjects are labeled "bad science" or "pseudoscience" or "anecdotal evidence" if applied to out-group subjects. Parapsychologists generally conduct their scientific studies more carefully that other scientists, because they are treated with suspicion. But it doesn't matter how carefully they conduct their experiments, if the scientific establishment has already decided (unscientifically, of course) that the subject matter does not exist.

The scientific in-group has its own vocabulary and mannerisms by which its members identify themselves to one another. I became aware of this when I was reminded by a Ph.D. scientist that scientists don't say "proof." They use instead the words "verification" and "falsification."

I remember that when I was first introduced to science in 1946, we were taught to pronounce the word "lever" as "lee-ver" and not "lev-ver." This is scientific in-group thinking and has nothing to do with the accuracy of science itself. These in-group preferences can change over time, like fashions in clothing. For all I know, "lee-ver" has now changed back to "lev-ver."

There is a whole vocabulary of words used in academic circles that I have to keep looking up in the dictionary, words such as "hubris" for "arrogance," "hegemony" for "domination," and "fatuous" for "stupid." The only reason I can see for using these unfamiliar words is that they serve as a kind of caste mark.

The in-group vocabulary and mannerisms are not "science." The real criterion that makes science valuable to the culture is its accuracy, and this accuracy is obtained by the scientific method. We need to look beyond the credentials and the in-group norms to see whether or not the opinions being expressed were arrived at by using the scientific method.

So we have uninformed opinions supported by unscientific methods accepted universally as absolute truth within the academic community. This should at least give rise to some debate. I thought that everything was debatable in the academic community. But there is no debate.

What forces are keeping everybody so compliant? Nobody is putting a gun to their heads. But, as I have already explained, the social forces of ridicule and ostracism threaten them with something very near to death. This threat easily explains their conformity.

Who is responsible for these forces? Who created these social laws that they all must obey? I have tried to explain it in terms of a social pecking order, with leaders and followers, but I don't see any obvious leaders. The illegitimate arguments of John B. Watson, Edward O. Wilson, and others I quote should be easy to spot by freshman students of critical thinking. So I can't imagine that people with Ph.D. degrees would be fooled by them. It seems that the people I quote are only resonating to a belief system already established. It is as if everybody in the academic community were held in the grip of some mass-hypnosis.

In the Dark Ages at least everybody knew who the Inquisition was and what power they represented. But in our time we are dealing with an invisible Inquisition representing an unknown power.

Because persons of the highest status in an in-group are allowed some degree of nonconformity, I sent letters in 2013 to the presidents or chancellors of 137 highest-ranked U.S. colleges and universities, asking them, as persons with the very highest status in the academic in-group, to use their influence to end the domination of physicalism in our educational institutions. I received 39 responses. They didn't all agree with me, but at least I put the idea in their heads. There are no obvious leaders enforcing the rule of physicalism, but maybe there are leaders who can end it.

Or maybe that is beyond their power. If the National Academy of Sciences or a consensus of Ivy-League presidents voted to end physicalism, it would happen. But if any one president, or an isolated few, made that suggestion, they would probably suffer the near-death experience of ridicule and ostracism. That is why I am now appealing to the general public, to put pressure on the academic community from the culture at large.

The members of the scientific establishment have done nothing in hundreds of years to eliminate the bias of physicalism. They don't seem to be even aware that it is a bias. Let us at least remind them that it is a bias and is seriously blocking both our cultural and scientific advancement.

CHAPTER 2

Scientific and Unscientific Methods

What is science? The *Encyclopedia Britannica* goes on for pages and pages on the subject of "Philosophy of Science," but I am trying to pinpoint only the essential attributes of science.

Science starts with observation. To have science, you must first observe the phenomenon being studied. This is in contrast to believing what other people tell you is true, even though they might be a priesthood claiming to know the absolute and ultimate truth, and even though they might be other scientists.

Our human perceptions do not give us an accurate representation of reality. They give us only a gross approximation of reality. They do not show us the disease germs. They do not show us the interactions of sub-atomic entities that manifest themselves as the reality that we perceive. They do not detect all the combined radio and television broadcasts that are in our environment constantly.

Given the limitations of human perceptions, science strives for the most accurate observations that can be made with these perceptions, and only for knowledge that can be obtained within these limitations. We have many kinds of instruments, such as microscopes and telescopes and particle-accelerators, to extend the reach of those perceptions. We can also extend the reach of our observations with logic and mathematics, by making logical inferences. But in no case does science make any assumption about anything beyond the reach of observation, even though it might be presented on the very highest authority.

Scientists don't always even believe other scientists. The second essential ingredient of science is replication. Scientists make their own observations and see if they agree with the observations of other scientists. Only if an observation is replicated many times over does it qualify as a scientific "finding." Replication helps to smooth out the errors that might be made in individual observations.

The third essential ingredient of science is accurate logic. I use the one word "logic" to mean the human ability to reason and its man-made extensions of formal logic and mathematics. The logic can be, and must be, 100% correct.

For a conclusion to be "valid," logically, it must follow from the observed conditions, and cannot be arrived at in any other way. Thus scientific experiments are designed to keep everything constant except the one variable to be tested.

But science doesn't have to be done in a laboratory. There are "field studies" as well as laboratory studies. Accurate observation alone, by trained observers, qualifies as "science." For example, hurricanes or animals in their natural environment can't be studied in a laboratory. But when enough observations of these things have been made, the criterion of replication is met, and a body of scientific knowledge can be developed.

So I see three essential ingredients in science — observation, replication, and valid logic — with observation, the first step, being the most important. Science can be observation alone, but to draw conclusions from that observation and call it a "scientific finding," "scientific determination," or even "scientific opinion," it must have all three ingredients.

I prefer to use the plural "scientific methods" rather than the singular "scientific method," because science takes many forms, with respect to the three essentials I have named above, and is not limited to any one procedure or approach.

James Bryant Conant of Harvard stressed the importance of theory in science, but theory is not a defining characteristic of science. Everybody has theories. We have conspiracy theories. Religious mythology is theory. But the defining characteristic of science is the system of observation and proof that backs up the theory.

The "unscientific methods" are all those arguments that are not scientific. And if they pose as "science," they are what I call "dirty science."

I don't want to try to describe all the unscientific arguments, because they are potentially infinite in number. I think that most college undergraduates are taught "critical thinking," where they learn the difference between legitimate and illegitimate arguments, starting with smear tactics and the classical "straw man," "ad hominem," and "red herring" arguments.

I define "smear words" as derogatory words or phrases without any supporting argument. "Straw man" means representing the opponent's argument as weaker than it actually is, so that it is easier to knock it down. "Ad hominem" means calling attention to flaws in the person, real or imagined, rather than in the person's argument. "Red herring" means to create a diversion, from the historical practice of dragging smoked herring across a person's trail to distract the bloodhounds from the scent.

And the list goes on from there. Instead of trying to identify all the items in an infinite list, I am going to cover only some of the methods that scientists are likely to use, and then point out as I go along, for each unscientific opinion that I discuss, why it is not scientific.

I actually saw and smelled a real red herring, or more accurately a bait box full of bright pink herring, the summer I was seventeen. The smell was overwhelming. It helped me to understand what I had not learned so well in the classroom, that creating a stink could be a major diversion. And it is helping me to remember now that I had begun learning about the illegitimate arguments in my prep-school English, long before I entered college. This education was continued in Harvard Freshman English, which was devoted largely to learning the difference between the legitimate and illegitimate arguments. In Social Sciences I, "Snarling Charlie" Cherington used to entertain us with mock propaganda, liberally seasoned with smear words, as a spoof of the kind of thing that Fox News does today.

The people I worked with in the computer field, I assumed, had all taken the same or equivalent courses, because illegitimate arguments were rarely used, and when they were used, management recognized them as such.

But it seems, in the scientific community, that once people earn their PhDs, they receive some kind of diplomatic immunity from being held to account for illegitimate arguments. They are free to call anybody "loony," or "fruitcake," or "crackpot."

Actually, the way this works is that they are highly accurate within their fields of expertise, and outside their fields of expertise, who cares? They are not held to account by their peers, who are largely in agreement with them anyway. So it is up to us common people to see, first of all, when scientists are offering opinions in areas where they are not qualified, and second, when they are using unscientific arguments.

The methods of status are the most common unscientific methods that scientists use. Status is what our parents had when we were small children and our parents were as gods to us, all-knowing and all-powerful. Status was what our teachers had when their word was Truth and our place was only to memorize. Status is what the clergy has, preaching the Word of God. We have all been conditioned in childhood to accept the authority of persons of status, not only to tell us what to do, but also to tell us what to believe. Persons of high status are the nobility, who say to the common person, "We are your 'betters.'" Status creates a caste system, separating people judged to be "superior" from those judged to be "inferior."

As science has achieved this kind of status, scientists are believed simply on Their authority. Any need for evidence is simply swept away. The scientific establishment simply becomes "Them," an all-knowing source of "Truth," and anybody so foolish as to disagree with what They say can be ridiculed and declared mentally incompetent.

Given this mental atmosphere, first the cultural conditioning that we have all had, and second the aura of omniscience that science has acquired, scientists can get away with making authoritarian pronouncements, not supported by scientific studies, expecting to be believed on the basis of their scientific credentials alone, and actually being believed on that basis. But without solid scientific evidence to back them up, such assertions are simply a bluff. Scientists are simply gambling that you wouldn't want to bet against their enormous status and credentials.

The most obvious of the bluffs are the statements of the form, "There are no such things" (UFOs, Bigfoot, spirit entities).

In order to make such a statement scientifically, one would need to have a complete knowledge of the entire universe.

The most extreme example of the authoritarian pronouncement is the scientist proclaiming, "You didn't see what you saw," usually regarding UFO sightings and alien encounters. First of all, the scientist wasn't there and couldn't have witnessed the non-event. Second, the scientist, unless telepathic, doesn't know what the person actually saw. And yet scientists actually get away with this stuff and convince a credulous public that the scientist's mind is so superior to that of the non-scientist that the scientist actually knows better what the person actually experienced, and that the person may actually have a mental problem.

Ridicule takes the bluff one step further, asserting that you are a fool for not knowing this Truth which is obvious to all scientists.

We need to call their bluff. We need to ask to see the scientific evidence.

I am counting on people who can think for themselves to see that the methods of science, and not the credentials of scientists, have given us our accurate cultural knowledge. And where scientists are trying to use their credentials to give credibility to unscientific methods, I am counting on you to call their bluff.

Refusal by scientists to look at the data is an extreme bias. When scientists refuse to look at the data, they cease to be scientists and become parental figures asserting their superior status. We need to remind them that by doing this, they have disqualified themselves as scientists. We need to throw the word "qualified" back in their faces to remind them that they who consider themselves "qualified" are actually not qualified.

Another source of extreme bias is the insistence that any physical explanation of a phenomenon is preferable to any non-physical explanation.

Be aware that as part of the bluff scientists may cite names of studies that don't exist or that don't apply. In this era of the Internet and smart phones, it should be easy to check them out.

Hard ridicule is more than just a bluff. Hard ridicule is persons in positions of power threatening the livelihood and the social existence of those people they are able to dominate. This has nothing to do with science. These are the methods of status,

methods of social manipulation and domination which have existed since the strongman ruled the primitive tribe with a club,
and before that, in the hierarchical "pecking orders" that exist in
animal species.

These methods of social manipulation and domination are
more powerful than the methods of science. One is not likely
even to express an interest in parapsychology, knowing that it
could be career-ending. These social pressures, unless actively
resisted, tend to take over, creating a bias, which diminishes the
accuracy of science. Persons of status in the scientific community
need to actively oppose the methods of status, to make sure that
scientific accuracy is not compromised by social pressures.

Lawyers who use illegal methods are disbarred. Similarly, scientists who use unscientific methods should lose their credentials. In any event, as an informed public becomes aware of what
these scientists are doing, they will lose their credibility.

Historically, persons of status have dominated by physical co-
ercion. Scientists, because they are more intelligent and better
educated than most of us, are more likely to dominate in a mental
way, with what I call "mental warfare," "mental combat," and
"mental bullying." By "warfare," I mean destructive actions
against other people through lies, deceit, and insinuations (as in
politics), as opposed to simply presenting the scientific evidence
with accurate logic.

Mental warfare includes all smear tactics. It includes what I
call "social pecking," or the kind of social derision used to work
on people's emotions, to manipulate them into conforming to
group norms. Mental warfare can actually kill people, as when it
causes teenagers to commit suicide because of social humiliation.
It includes the kind of false and damaging statements that are
defined legally as "defamation," but it is not limited to any legal
definition or even to things that people actually say.

Mental warfare has been characterized vividly as "shark attacks" by Larry Burk in his book, *Let Magic Happen* (Burk, 2012,
pages 155-160). This comes from a paper presented by Richard J.
Johns (Johns, 1987), allegedly translated from a paper in French
by Voltaire Cousteau, who died in 1812. This clever spoof claims
to have been written for sponge divers, but is really aimed at the
academic world. The object of shark attacks is not to establish

any scientific truth, but simply to rip apart any nonconforming argument by any means.

I have observed that some people with powerful mentalities are actually able to block my thought processes while I am in their force field. I call this "mind-jamming" and "mind-scrambling." We need to be aware that this is a possibility when we are going head to head with the powerful minds of scientists. The "shark attacks" probably incorporate this ability, plus any other tricks that a powerful mind can produce. (Members of the scientific establishment will ridicule the idea of a mental force field at the same time as they are using it.)

Warfare wins wars, but warfare does not create accurate knowledge.

In smear attacks, there is a Speaker, a Target, and an Audience. The Speaker says derogatory things about the Target, to be believed by the Audience. This involves some mental phenomenon related to hypnosis, to implant the suggestion in the minds of the Audience that the smear actually applies to the Target. The attention of the Audience is focused on the Target. The attention of the Audience should instead be focused on the Speaker, as the person creating the smear, and as the kind of person who creates smear tactics.

Richard Milton, author of *Shattering the Myths of Darwinism*, successfully turns this around when he points the finger at critic Richard Dawkins:

> ... according to a review by Darwinist Richard Dawkins, the book is "loony," "stupid," "drivel" and its author a "harmless fruitcake" who "needs psychiatric help."
> (Milton, 1997, page ix)

When I read this, I asked myself, "Don't they have laws against defamation in England?" Thus I was prompted to take a trip to the Duke University Libraries, to find out what Dawkins really said.

Locating the review required a great deal of help from the librarians, because the *New Statesman* from 1992, which Milton cited, went bankrupt and did not exist in 1992. Instead, we found the review in something called *New Statesman & Society*. It was on microfilm and barely legible, but worth finding, because it showed how Dawkins was clever enough to do his defamation without actually saying that Milton was all those things.

He starts off by saying:

> Every day I get letters, in capitals and obsessively underlined if not actually in green ink, from flat-earthers, young-earthers, perpetual-motion merchants, astrologers and other harmless fruitcakes. The only difference here is that Richard Milton has managed to get his stuff published.
> (Dawkins, 1992)

In a very roundabout way, by criticizing Milton's publisher, Dawkins manages to convey to the reader that Milton belongs in the category of people he calls "harmless fruitcakes." As for "needs psychiatric help," he quotes Francis Crick as saying, "Anyone who believes that the earth is less than 10,000 years old needs psychiatric help," thereby avoiding again the direct defamation of Milton.

Actually, Milton is questioning the dating methods used by scientists and the accuracy of the thought process upon which those methods are based. Dawkins ignores that whole argument, instead saying, "Perhaps the world really did bounce into existence in 6006 [sic] BC," implanting the idea in our heads that Milton is using the fundamentalist argument, without actually saying that he does. So we can't really call Dawkins guilty of misrepresentation here, either.

He spices up the review with a generous collection of smear words, starting with the title, "Fossil fool," and followed by "obsessively," "fruitcakes," "silly," "fabrication," "unqualified hack," "fairies," "werewolves," "loony," "wretched publisher," "egregious howlers," "twaddle," "complete and total ignorance," "nonsense," "stupid," and "silly-season drivel," without successfully applying these descriptions to Milton or his book, but just firing off these shots to enhance the flavor of the review.

This is the first of many lists of smear words that I will offer, in order to identify the poison in what some people are saying. I offer it as an exercise for students of critical thinking to go through a piece of writing and pick out and underline the obvious smear words, such as "silly," "moronic," and "idiotic," and then ask whether the whole piece of writing isn't just a setting for these gems and their intended purpose of manipulating people emotionally.

Dawkins is particularly expert at this kind of manipulation, and Milton does a good job of turning things around, pointing out the unscientific nature of the criticism by saying, "This is not the language of a responsible scientist and teacher." (Actually, Dawkins is so manipulative in this review that I don't even believe his biology.) Milton, the Target, has pointed the finger back at Dawkins, the Speaker, to show that Dawkins's unscientific language more accurately represents Dawkins and not Milton.

Milton then goes on to say that Dawkins also attacked his publisher for being so irresponsible as to publish a book criticizing Darwinism. This can be turned around, too. The *New Statesman*, which advertises "the intelligence, range and quality of its writing," escapes because of its temporary condition of bankruptcy, but its placeholder, the *New Statesman & Society*, can certainly be criticized for publishing anything as irresponsible and unscientific as Dawkins's review.

The Speaker accusing the Target of doing something that the Speaker is actually doing is psychological projection. It serves the double purpose of directing attention away from what the Speaker is doing, at the same time that it is smearing the Target. This is done all the time in politics. I remember once in New Hampshire receiving Republican junk mail warning me of "Democratic dirty tricks." The Democrat, Katrina Swett, conducted a pristine campaign, while the Republican, Charlie Bass, waged his usual dirty campaign.

Hypnosis is ridiculed by the scientific establishment. And yet something related to hypnosis is used all the time in our daily lives. Every time anybody makes a statement, that statement is first implanted in our minds and memories, as in a recording, before we are able to examine it critically. If we do not examine it critically, it becomes part of our belief system. Even if we do examine it critically and reject it, it remains recorded in our memories (conscious or subconscious) forever.

When the Speaker makes a statement about the Target, it is implanted in our minds as a representation of the Target. It is necessary to be aware of this consciously, and recognize that every statement is a reflection of the Speaker and not the Target. If the Speaker uses smear tactics, it should be recognized that the

Speaker is the kind of person who uses smear tactics, and the Speaker should lose credibility as a result.

Another ingredient of "shark attacks" is what I call "preemptive misrepresentation." Whatever you DON'T say can and will be used against you. If you haven't told these people your whole life's story, they can jump in with some quick fiction to discredit you, in some area that you haven't covered. For example, they might call me a "believer" because I haven't yet said where I stand on the subject of religion. And they have probably already dismissed the dream of my grandmother's death with some quick fiction, to secure their belief that such things are impossible.

Watch out for implied premises. If you let them use the word "supernatural," they have half-won the argument already, because it implies that what is "natural" has already been defined and they know what it is.

Also the word "skeptic" is used to convey the false premise that a person is open-minded, when actually these people are zealots whose minds are totally closed to anything beyond establishment views. They are more appropriately known as "pseudo-skeptics." They will sit there with a smile on their face and find everything you are saying totally amusing.

The shark attack experienced by Larry Burk started out with the gambit, "This course needs to be taught by a skeptic." Immediately the false premise has been inserted into the discussion, and then, without allowing time to determine what is meant by "skeptic," or whether the course should appropriately be taught by such a person, the quick fiction continues with, "Dr. Burk, do you consider yourself a skeptic?" Faced with this interrogation, Burk says, he felt like the proverbial deer in the headlights (Burk, 2012, page 156).

Other examples of unscientific methods I will give as I go along, as they apply.

Politicians can use invalid arguments, dirty tricks, attack ads, and opponent-bashing, in order to win votes. But scientists can be held to a standard, the scientific method. If they are using unscientific methods, it is not necessary to defeat them in warfare; it is necessary only to point out that they are not doing science and are losing credibility as a result.

CHAPTER 3

The List

The academic establishment, like the scientific establishment, is an in-group, with a long list of opinions, attitudes, and beliefs, which, like physicalism, are not supported by legitimate arguments, and yet, like physicalism, are implanted in the minds of academic people so completely that debate is out of the question. These ideas are just parroted by academic people, as if they represented fundamental truths that are self-evident and need no supporting arguments. Here is a list of the ones that I am aware of:

"The myth of the self-made individual"
"Freud did bad science."
"Carl Jung was a mystic."
"Ayn Rand was right wing."
Edgar Cayce (they insinuate) was a fraud.
Eklal Kueshana is thought not even to exist.
Creativity doesn't exist.
Intuition doesn't exist.
Psychic abilities don't exist.
Parapsychology is pseudoscience.
The spiritual doesn't exist.
"Dreams are a random firing of neurons."
"The mind is nothing but the physical brain."
prejudice against football
prejudice against psychotherapy
prejudice against hypnosis
prejudice against religion

These things were communicated to me by scientists and scholars at the very highest level, including a Nobel Prize winner and a Pulitzer Prize winner. I can't believe that people at this level would take such prejudices and beliefs seriously, let alone allow their thinking to be ruled by them. So again the possibility of mass hypnosis comes to mind.

I use the expression "hypnotic implant" because the word "suggestion" isn't strong enough. As I have said, a hypnotic suggestion becomes implanted in one's mind before one's critical faculties are able to deal with it. Therefore I call it a "hypnotic implant." Or one might call it a "hypnotic command," because, once implanted, it can operate as a law that must be obeyed. I offer the hypothesis, to be questioned, to be tested, to be debated, that every item on The List operates as a hypnotic implant or command. These hypnotic implants are not rejected by people's rational thinking because the academic community, like the scientific community, is an in-group, and these hypnotic implants are sustained by constant reinforcement from one's colleagues, who are programmed in the same way. I offer my rational arguments in this book as a way to break out of this spell, and I welcome the rational arguments of academic people in response.

In September 1952, the day before we began classes at Harvard, they packed our entire freshman class into Sanders Theatre and gave us official permission to use our own minds, to think for ourselves. This was a great day for me, after many years of authoritarian schooling. It was a permission that not every college student enjoyed, as I learned in later life from a friend who had gone to Georgia Tech. It is called "academic freedom," or "freedom of inquiry." This freedom is important for the acquisition of new knowledge, but it is severely restricted by the biases on The List. It becomes freedom to think as They think. I began to understand this while I was at Harvard. People who were more savvy than I was got good grades by learning the biases of their professors. Yes, individual professors had individual biases, which might affect one's grade, but I had no idea that there were universal biases, across the board, that affected the entire academic community, restricted academic freedom and freedom of inquiry, and could destroy one's whole social existence.

After graduating from Harvard, I broke away completely from the academic community. I saw the threat of nuclear annihilation as proof of total systems failure, not of the technology, but of the culture that had created it. Because of this and other cultural problems, I turned my back on the culture and the academic community which represented the culture, and set out to design a new civilization — to figure it all out for myself, from the beginning. Of course I had the benefit of the cultural knowledge, but not without questioning it, starting with fundamentals, such as "Can I use logic?" I decided for myself what was true and false, without asking the academic community to give me a grade. And of course, without really being aware of it, I also freed myself from the prejudices of the academic community.

And then, when I presented my developed system to the academic community, asking them to evaluate it, I was met with a deafening silence. They didn't tell me what the problem was. I had to figure it out for myself. Innocently, unknowingly, I had violated many of their taboos.

Or, if you turn that around, you can see that in my innocence I was able to go ahead and design a system that the academic community was prevented from creating because of their taboos. If the academic community was aware that their thinking was limited by taboos, and how much our culture has been crippled by those limitations, they would recognize the extreme value of simply being able to break free of that cage and think without those restrictions. That would be a major accomplishment in itself, whether or not such a person was able to design a great system. Credit the Hippies. Credit a great many people in New Age pursuits. But the academic community, which defines and represents our culture, is not able to see all that, because they are still locked in the cage. So how can they break free of this cage? I think the first step is to become consciously aware, and to admit, that these prejudices exist.

I realize that turning my back on the culture runs contrary to the legitimate values of academic people. Even so, I think they might have stayed with me to see what this approach might produce, if it hadn't led me into taboo areas.

The prejudices against football and Ayn Rand did not affect me. I include them here only because they exist. As much as I enjoyed reading *Atlas Shrugged*, it didn't have the answers I was looking for, because it ended up with the world in flames. But Ayn Rand serves as a good simple example of the kind of erroneous thinking these prejudices are built upon.

Ayn Rand was not "right wing." "Right wing" means wanting to go back to the past. She described herself as a "radical capitalist." She showed, in *Atlas Shrugged*, in 1957, what Gorbachev was to proclaim 30 years later: "Socialism doesn't work." She was trying to go beyond socialism, to the future, not back to the past, to design a system that would work. But I suppose to the academic establishment, with their bias towards socialism, it is hard to see the difference between being a "capitalist" and being "right wing."

The first item on The List, "The myth of the self-made individual," would block my work completely. It makes a "straw man" out of the so-called "self-made man," saying that he didn't invent the language or write all the books, and therefore wasn't entirely self-made. But actually, all the self-made man is claiming is that he received his education, or a large part of it, outside of the formal institutions of education.

When I say I am "Harvard-educated," I am not claiming that Harvard invented the language or wrote all the books, but only that Harvard determined the course of study I was to follow. And similarly, when I say I "re-educated myself," it doesn't mean that I invented the language or wrote all the books, but only that I determined the course of study I was to follow, after Harvard, in my quest for a new civilization.

And that course of study involved almost every no-no on The List: Freud, Jung, psychotherapy, dreams as education, Edgar Cayce, and Eklal Kueshana. I developed my creativity, intuition, and will — mental attributes thought not to exist. I had precognitive dreams. I literally "saw The Light." And in my innocence I violated a dozen irrational taboos.

Or should I call them "sacred slogans" or "shared delusions?" I don't have words strong enough to describe the impact these beliefs and prejudices and hypnotic implants have had on our culture and its hopes for advancement. Students looking for a liberal education and "freedom of inquiry" are instead being restricted by a quasi-religious cult mentality.

"The myth of the self-made individual" needs one more comment: By closing its eyes to the "self-made individual," the academic community also closes its eyes to what that individual may have learned outside the confines of academia. The head of Harvard University Press told me in 1970, "We publish books by scholars for scholars." The "by scholars" part implies people with Ph.D. degrees operating within a "school." It implies that the academic community is recirculating the same information and is closed to new knowledge coming in from the outside world, from people who aren't "scholars."

There are scholars, and there are creative thinkers. The scholars preserve and teach our cultural knowledge. The creative thinkers create new knowledge. Usually the creative thinkers are also scholars, but this is not always the case. I am a creative thinker, but not a scholar. I concede that the scholars have much more knowledge than I have. But also I have knowledge the scholars don't have, because I created it. It is this knowledge that I am trying to communicate to the scholars, for the advancement of the culture. When the members of the academic in-group recognize the limitations placed on their own thinking by in-group pressures, they will value more the creative thinking coming to them from the outside.

The cost of attending Princeton, Harvard, Yale, or Stanford (PHYS) for the academic year 2018-19 is almost $70,000. For this price, I think prospective students should ask what they are getting for their money.

The argument that obscures this issue is economic: If you have a college degree, your lifetime earnings will likely be much more than what you would have earned with only a high-school education, and the return on your investment will likely be many times what you initially paid for your college education. I say "likely," because there are exceptions, such as Steve Jobs and Bill Gates, but for most people a college education is a good investment, and an education at one of the most prestigious institutions is an even better investment.

This argument of course bypasses the whole issue of the quality of education. That quality is compromised as students are indoctrinated into a limited view of reality. They aren't encouraged to develop their full human potential. The economic equation can change as companies recognize the value of things such as psychotherapy, thinking outside the box, and the power of prayer in their employees, not to mention psychic abilities.

Why aren't there alternative universities that teach this stuff? There are, but for the most part they are ranked lower than the lowest ranked university, probably because of the same prejudices that created The List in the first place.

I am going to pick apart most of the items on The List in the course of this book, showing how they reinforce physicalism in blocking our view of the mental and the spiritual. But the legitimate arguments are not enough. The academic establishment has power, power that does not come from the legitimacy of their arguments, but power that comes from the fact that a college education is a good investment, no matter what was taught, and power that comes from the fact that scientists are believed as "authorities," no matter what they are saying.

In our government we have "checks and balances" to keep any one individual or faction from having absolute power. In business, competition helps to keep people honest. But the academic establishment as one in-group is in a monopoly position, with no established checks and balances.

So I need your help in opposing this power. I need you to question their authority every time they parrot one of the slogans from The List. I need you to ask them what academic department is responsible for knowing about Carl Jung, Edgar Cayce, Richard Kieninger, reincarnation, levitation, precognition, or spirit communication — or for having demonstrated rigorously that none of these subjects are worthy of study. I need you to engage them in debate.

And most important, I need you to refuse to support your alma mater if it is dominated by these unscientific and unscholarly beliefs. I have not given money to Harvard since 1966, because Harvard supports "scientism," as I called it in my 45th Reunion Report. And I have told them about it over and over again. A few big donors withholding their money and TELLING their universities why they are withholding it should have an effect on the power equation.

Or, better yet, the research that Ian Stevenson was able to do on reincarnation at the University of Virginia was made possible by a large donation specifically for spiritual and religious studies. So you could apply this financial power in a positive way, by specifying that your donation is for the study of the spiritual, or parapsychology, or the siddhis.

Money is power. Money determines, to a large degree, what academic people study. The people with the money should not be affected by the ridicule, and certainly not by threats of loss of funding. So I hope you will read my legitimate arguments here and translate them into power with your financial support or non-support of the academic community.

CHAPTER 4

Reflections of Academia

Harvard biologist Edward O. Wilson has won two Pulitzer Prizes. Classmates of mine have spoken very highly of him. He was the honored speaker at our 50th reunion. But when he ventures opinions outside of his particular field of expertise, he doesn't do so well, in my opinion.

In his book *Consilience* (1998), to the best of my understanding, he is saying that if we all thought like biologists, we would have "consilience," or agreement, in the world. To that end, he dismisses or puts down other ways of thinking. I may be oversimplifying it, but that's the way that I see it.

What makes *Consilience* interesting to me is that it seems to be reflecting some of the opinions, attitudes, and beliefs of the social group academia at the time. It also displays some of the unscientific methods.

Let's start off with some social manipulation:

> I mean no disrespect when I say that prescientific people, regardless of their innate genius, could never guess the nature of physical reality beyond the tiny sphere attainable by unaided common sense. Nothing else ever worked, no exercise from myth, revelation, art, trance, or any other conceivable means; and notwithstanding the emotional satisfaction it gives, mysticism, the strongest prescientific probe into the unknown, has yielded zero. (Wilson, 1998, page 46)

First of all, when he says, "I mean no disrespect," he means that he does mean disrespect and is making an effort to disguise it. He then proceeds with his put-down of the "prescientific," whatever that means.

40

As a scientist, he should know the logical implications of saying that anything has yielded "zero." This is just an authoritarian pronouncement, taking advantage of his enormous status as a scientist.

Also, the placement of the word "zero" at the end has the hypnotic effect of implanting the "zero" in our minds and ending the discussion.

With his enormous credentials as a scientist, he describes five "diagnostic features of science that distinguish it from pseudoscience":

1. repeatability (replicability)
2. economy (elegance)
3. mensuration (measurability)
4. heuristics ("stimulates further discovery ... and the new knowledge provides an additional test of the original principles ...")
5. consilience (consistency with explanations of other phenomena)

And then he goes on to say:

> Astronomy, biomedicine, and physiological psychology possess all these criteria. Astrology, ufology, creation science, and Christian Science, sadly, possess none. ... (Wilson, 1998, page 54)

With the one word "sadly" he sweeps away all these things that he calls "pseudosciences" — before we have even had a chance to think about what he is saying. "Sadly" is the kind of word that publishers use when they reject your manuscript. It implies that somebody has the absolute authority to make a judgment, and that that judgment has already been made, and that now the only thing left to do is to tell you the sad news.

In the face of Wilson's enormous scientific credentials, I question his authoritarian pronouncement right from the beginning. Science is accurate observation and valid logic applied to phenomena of the real world to create knowledge. I don't see the need to pile upon it all these intellectualizations, some of which I see as essential to science (replicability) and some as not.

And then I question his "sadly" in regard to Christian Science, which has:

1. Repeatability: Mary Baker Eddy called it "science" because people sent her testimonials of miraculous cures, thousands of them.

2. Economy: The theory is simple enough: You can use prayer to combat disease. Also certain mental attitudes, such as not believing in disease, are effective.

3. Mensuration: You live or you die. You get well, or you don't. You do better than the doctors expected, or you don't.

4. Heuristics: People have devised experiments to test these theories and expand our knowledge in this area (Pennebaker, IONS).

5. Consilience: The Catholic Church also has thousands of records of miraculous cures. Many of Edgar Cayce's cases were cases in which medical science had failed. Also the theory is consistent with the Edgar Cayce readings.

(Much of this stuff that so-called "science" is rejecting is becoming a consistent body of knowledge in itself.)

Where Christian Science is incomplete as science is that it doesn't have a logically complete picture: How often has Christian Science failed to overcome disease, and how often have there been spontaneous "miraculous" cures, without the application of Christian Science? We don't know, but certainly there are things that are called "scientific" that do not have a logically complete picture, either. And certainly it is not accurate to say that Christian Science has "none" of the attributes of science that Wilson describes.

In his blanket dismissal, Wilson lumps in Christian Science with other subjects which are unrelated. Each subject deserves its own individual treatment. Creation science is not science. Astrology has been tested scientifically, but the bias is so strong on both sides that the believers and disbelievers both find the results that they want. Ufology seems to be operating in the face of an apparent government cover-up. But all these things do share one thing in common, to continue with Wilson's statement:

... And it should not go unnoticed that the true natural sciences lock together in theory and evidence to form the ineradicable technical base of modern civilization. The pseudosciences satisfy personal psychological

needs, for reasons I will explain later, but lack the ideas or the means to contribute to the technical base.
(Wilson, 1998, page 54)

All of these things he calls "pseudosciences" are threats to what he calls "the ineradicable technical base of modern civilization." They all indicate that there are more dimensions to reality than what we now know as physical reality. The technical base we now have is incomplete, and needs to be opened up to include other phenomena, if our knowledge is to expand — to keep us from eradicating it and us. He is arguing his point not with a valid scientific argument, but by suggesting that people who are trying to extend our knowledge further have "personal psychological needs."

Actually, the "personal psychological needs" work the other way. There is a certain satisfaction and security in thinking that you have most of the answers, especially when you are being paid to do just that. Within this comfortable security, there is a resistance to see or admit that whole dimensions of the universe have barely been explored. This knowledge-inertia keeps belief systems in place. It is the scientific establishment that has the "personal psychological needs." Wilson is giving us here an example of psychological projection, where the Speaker is accusing the Target of something the Speaker is actually doing himself.

On page 74, there is a quick smear of Swedenborg: "I suspect that he would have enjoyed a stiff dose of *ayahuasca*."

Wilson then proceeds to contribute to the Freud-bashing. He sets the tone by comparing dream symbols to "characters in a bad Victorian novel," thus giving people the social signal that Freud is being ridiculed. After a reasonably accurate, although colored, description of Freudian psychology, he then proceeds with his authoritarian pronouncements:

> Freud's conception of the unconscious, by focusing attention on hidden irrational processes of the brain, was a fundamental contribution to culture. It became a wellspring of ideas flowing from psychology into the humanities. But it is mostly wrong. Freud's fatal error was his abiding reluctance to test his own theories — to stand them up against competing explanations — then revise them to accommodate controverting facts. ... In

dreams Freud was faced with a far more complex and intractable set of elements than genes, and — to put it as kindly as possible — he guessed
wrong. (Wilson, 1998, page 75)

"But it is mostly wrong." Again, does he have enough status
to pull off this demolition of Freud, or is he going to lose his
own credibility as a result? And the punch line, "to put it as kindly as possible," is really very unkind, as an attempt to gain the
sympathy of the reader as he bashes Freud one more time.

If he wants some credibility, he needs to make a list of Freud's
errors, show how they outweigh Freud's discoveries, and explain
how any errors are "fatal." Without any details, he loses this argument by default.

Actually, Freud was mostly right, although he did make a
number of errors. None of these errors were fatal, and they were
corrected by Carl Jung and others, including myself. Procedural
errors have been noted by physical scientists, but physical scientists, including Edward O. Wilson, are simply not qualified to operate in the area of the mental. Freud's theories have been thor-
oughly tested by me and millions of other people in psychotherapy, using the mental senses.

Freudian psychology has now gone well beyond Freud. People
who are trying to discredit psychoanalysis by bashing Freud are
missing their target by at least 70 years.

The culture accepted Freud completely until 1973. Freud was
"in;" then Freud was "out." I think part of the reason for the
Freud-bashing is emotional, to compensate for having accepted
him too completely in the past.

When Wilson tries to describe dreams, he is largely uninformed,
reflecting the opinions, attitudes, and beliefs of physical scientists
of his time, based on physical images of the brain at work:

In brief, dreaming is a kind of insanity, a rush of visions, largely uncon-
nected to reality, emotion-charged and symbol-drenched, arbitrary in content, and potentially infinite in variety. Dreaming is very likely a side effect
of the reorganization and editing of information in the memory banks of
the brain. It is not, as Freud envisioned, the result of savage emotions and
hidden memories that slip past the brain's censor. (Wilson, 1998, page 75)

Again, here is the authoritarian pronouncement.

I am trying to find an appropriate metaphor to explain why the brain-scanning equipment is not able to experience a dream as a person does, using the mental senses. Let's say you are completely deaf, and you are trying to experience a violin concerto by physically monitoring the vibrations of the violin. Yes, this brain-scanning equipment is truly awesome, but that's about as close as it is going to get to understanding dreams. There is just no connection between the physical monitoring of brain waves and determining whether Freud and Jung were wrong.

Dreaming is triggered when acetylcholine nerve cells in the brain stem begin to fire wildly ... (Wilson, 1998, page 76)

What causes these cells to begin to fire wildly?

... In dreams we are insane. We wander across our limitless dreamscapes as madmen. (Wilson, 1998, page 77)

This has propaganda value, to portray our dream states as insanity. But I know differently. I have experienced a meaningful and purposeful education from my dreams, as valuable as my four years at Harvard. I have experienced "the self-steering process," guiding me towards the truth. So I know that Wilson doesn't know what he is talking about here. Again, he is just picking up on ideas that are floating around academia.

The brain and its satellite glands have now been probed to the point where no particular site remains that can reasonably be supposed to harbor a nonphysical mind. (Wilson, 1998, page 99)

Is a nonphysical mind supposed to exist in a physical place? This is the same argument as probing the physical universe and finding no God. Other dimensions need to be explored.

On the subject of evolution, Wilson says:

... Perhaps God did create all organisms, including human beings, in finished form, in one stroke, and maybe it all happened several thousand years ago. ... (Wilson, 1998, page 129)

Like most people who believe in evolution by purely physical means, he is picking on the Fundamentalists. It is easy for them to pick on the Fundamentalists, who aren't generally as intelligent or well educated, and are tied to a rigid belief in a text that was written thousands of years ago. The chronology of the Creation of course has been totally disproved scientifically.

It is also easy to pretend that the only opposition to the idea of physical evolution is coming from the Fundamentalists, and that it is necessary only to show that the world wasn't created in 6 days in 4004 BC. Later on I'll be presenting Eklal Kueshana's view of evolution, which fits the evidence better than Darwin's theory does.

In all these things — dismissing Christian Science, UFOs, astrology, and the spiritual, bashing Freud, seeing dreams as meaningless nonsense, and accepting the purely physical view of evolution — Edward O. Wilson is reflecting views of the academic in-group that were fashionable at the time.

CHAPTER 5

Unintelligent Design:
From Theory to Dogma to
Cognitive Dissonance

The key to the belief in a purely physical universe is the belief that life simply evolved without any intelligent design — that matter simply assembled itself according to laws of chemistry and physics. Thus there is no need to imagine a Creator as an explanation of how living things came to exist.

Unintelligent design is plausible in the case of inanimate matter. Hydrogen atoms, through nuclear reactions such as the ones right now taking place in the sun, could have fused into helium atoms, and the helium atoms could have fused to create heavier atoms, which then became random materials scattered throughout the universe. Clouds of random materials would be condensed into spherical objects by the force of gravity. Within those spherical objects, further atomic, physical, and chemical interactions would take place. We would have a pattern of mineral substances deposited as we now see them in the universe, with clear physical explanations of how they became that way.

But living things are much more complex than simple mineral substances appearing here and there. A living thing right away is organized to do something. A living thing interacts with the materials in its environment to create other organized structures similar to itself. Right away a self-replicating entity is far more complex than anything that human beings have ever invented. (No, a self-copying computer virus doesn't count. A computer/robot that is able to go out into the real world and find the raw materials and make the parts and assemble them into an entity like itself is more like it.) I am appealing to the medical profession's understanding of just how complex a system the human body is, and what a complex code of genetic information is built into every cell. Of course life didn't start this way. It started with simple one-celled organisms. But they all had the capacity to create more of their kind from the material in their environment. And, I repeat on purpose, no human being has ever (to my knowledge) created such a self-replicating mechanism.

So we had to have the concept of a "God" to explain such a thing. Actually, we had to have the concept of a "God" to explain the creation of the universe as well. Scientists now have the "big bang" theory, but that doesn't explain it. It only begs the question: What created the big bang? There are questions we can't answer. Philosophers and scientists should be smart enough to know that we can't answer them. But still people try.

Charles Darwin observed that various species were able to modify themselves, through random mutation and natural selection, to adapt better to their environment. Offspring are never exactly the same as their parents (this is the "random mutation" part), and those offspring that are better adapted to their environment have a better chance of surviving through their reproductive age and passing on their particular adaptive features to their offspring (this is the "natural selection" part). This whole process of species being modified by random mutation and natural selection, to better adapt to their environment, is called "adaptation."

Adaptation has been proved many times by the biologists, and can be demonstrated mathematically. If a certain mutation enables an individual to survive and produce 10% more offspring than an individual without the mutation, then after one generation, that individual will create 1.1 times as many offspring as an individual without the mutation. If the mutation is also passed along to the offspring, then the second generation will gain another 10% (1.1 times 1.1) and have 1.21 times as many offspring as those without the mutation. It is an exponential progression, with the proportion increasing by 1.1 times with each generation. So after 128 generations, the descendants of an individual having the particular characteristic will outnumber the descendants of another individual whose descendants have never been interbred with that characteristic by 1.1^{128} times, or about 200,000 to 1.

From his observations, Darwin theorized that the process of adaptation by which species were able to modify themselves was also the process by which new species were created. Apparently he wasn't all that sure of his theory, because he waited 20 years to publish it, until Alfred Russel Wallace proposed a similar theory, thus pressuring Darwin to publish his own. Then T.H. Huxley and others picked up on it and extended it into a theory of the origin of life itself.

This extension of Darwin's theory made physicalism a plausible belief system. We didn't need a God, a Creator, as an explanation of how living things came to exist. It was possible to believe in a purely physical universe. People who had outgrown the need to believe in the mythology of an authoritarian and judgmental religion seized the opportunity to get rid of the spiritual altogether and believe only in the existence of the physical reality.

I have been in trouble calling Darwin's theory "The Theory of Evolution," because then people switch definitions on me and point to the "mountain of evidence" supporting the evolution of life. First of all, this definition-switching is an illegitimate argument. Scientists know this, and even have a word for it. But because of those people who might want to switch definitions on me, I no longer use the word "evolution" to describe Darwin's theory. I refer to it simply as "Darwinism." And the idea that life evolved without benefit of any intelligent design, I call simply "unintelligent design." I don't want to argue about what Darwin said or didn't say. I want to focus only upon the essential point of whether life evolved without benefit of intelligent design. That theory was not wholly Darwin's, but evolved out of the cultural consciousness of the time. But Darwin's theory, or "Darwinism," was the key to the belief in unintelligent design, which in turn was the key to the belief in a purely physical universe. And the idea of unintelligent design was adopted by the culture and was fully embedded in the cultural consciousness long before the evidence to support it had been accumulated.

And the evidence, once accumulated, did not support it. In 1983, I read in the *Encyclopedia Britannica* that new species have appeared too quickly in geological time to be accounted for by the process of adaptation, which is too slow. It seemed, to me, that this disproved Darwin's theory. And given the huge complexity of any living species and the relative suddenness at which new species appeared, it certainly seemed that some kind of intelligence was at work making this happen. The evidence was more in favor of intelligent design than unintelligent design.

But the biologists were blocked by the dictates of physicalism from offering any such so-called "supernatural" explanation. They were allowed only a physical explanation. Some of them simply denied the evidence. Others kept Darwin's theory alive by modifying it with the theory of "punctuated equilibria," hypothesizing that some radical change in the environment must have forced the process of adaptation to speed up.

The problem with that theory is that you have to show some radical change in the environment for every new species that evolved too quickly.

Or, as was pointed out by a person I knew who had studied biology, the most likely result of a radical change in the environment would be extinction.

More recently I have become aware of the "Cambrian explosion," some 450 million years ago, when many new species appeared suddenly (in geological time) fully formed, with very little evidence of the gradual steps of adaptation. This is more evidence to refute the theory that the process of adaptation was the same process that produced new species.

"Adaptation," the modification of individuals within a species, and "speciation," the evolution of a whole new species, are two different things. Animal breeders know that there are limits beyond which members of a species cannot be modified. As far as I know, speciation has never been accomplished in the laboratory, whereas adaptation has been achieved many times.

Where did I get this information? I got it from the biologists. It appears that the biologists themselves, in amassing their "mountain of evidence" over the years, have actually found the evidence to refute Darwin's theory. But they can't admit that, because Darwin's theory is the key to their whole belief system. In defense of their beliefs, they come up with authoritarian statements assuring us that Darwin's theory is still alive and well.

At some point, Darwin's theory ceased to be a scientific hypothesis and became quasi-religious dogma. Around 1950, the biologists were searching for "the missing link." They never found the missing link (which would have shown the continuity of the adaptation process), and then they simply stopped searching for it. By the 1980s, when I tuned in to Stephen Jay Gould on TV to see what proof of Darwin's theory he could offer, he simply started off with Darwin's theory as a "given," a first premise, a fundamental truth, which he built upon by showing us a wonderful example of adaptation.

The biologists are suffering from what Leon Festinger called "cognitive dissonance." This is when serious believers suddenly find their beliefs contradicted by evidence. They could use some psychotherapy, but their in-group would shun them if they turned to psychotherapy, and especially if they abandoned physicalism. They are stuck.

So they turn their intelligent minds to clever arguments. When Phillip Johnson published *Darwin on Trial*, the physicalists turned it around and put God on trial, demanding that he prove that God exists. And they were ridiculing "intelligent design," it seemed, even before anybody seriously proposed it.

And they parade their credentials. T.H. Huxley ridiculed people who had less of an education than he had. In answering the critics of his essay "On the Physical Basis of Life," he said:

> ... My unlucky "Lay Sermon" has been attacked by microscopists, ignorant alike of Biology and Philosophy; by philosophers, not very learned in either Biology or Microscopy; ... (Huxley, 1871, page vii)

More than a hundred years later, Richard Dawkins continues along the same lines:

> Darwinism, unlike "Einsteinism," seems to be regarded as fair game for critics with any degree of ignorance. (Dawkins, 1987, page xi)

"Einsteinism," of course, is accepted because it has been proved for all the world to see, at Alamogordo, Hiroshima, Nagasaki, and a thousand other places. You don't have to know the meaning of "$E = mc^2$" to know that it is the power that ended World War II, and the power that has since threatened to annihilate us.

Darwinism, on the other hand, is far from proved, and after eight hours reading about "Evolution" and related subjects in the *Encyclopedia Britannica,* I was able to see that the evidence was actually against it. So yes, the weak points in the evolution argument can be easily pointed out, accurately, by "critics with any degree of ignorance," like me.

Much has been written about "natural selection," but what about that other part of the theory? What is meant by "random mutation?"

We have this word "random" to indicate that we don't know exactly what the outcome will be. We can't calculate mathematically which sides of the dice will end up up, but we can say that each one of the six sides has equal probability of being up, and do the math from there. We can't calculate exactly where a leaf is going to end up, falling from a tree, but we can make some statement about the mean and standard deviation, given no wind, of course.

And we know that there are small differences among individuals of the same species, although the basic structure remains the same, but we don't know exactly what forces cause these differences. And until we know what forces cause these differences, we don't know for certain that living creatures evolved by means of purely physical forces.

It is not necessary to postulate a God, or Divine Intervention, or even an intelligence creating new species. It could simply be rules of chemistry and physics that we don't know about yet. But we really don't know the process by which new species evolve. And the quickness by which new species evolve, as compared to the slowness of adaptation, suggests that these are two different processes.

The implied premise in Darwin's theory is that new species evolved WITHOUT ANY DIVINE INTERVENTION. Here is where biologists are stretched beyond the limits of biology by Darwinism. Because their observations are limited to those of the physical senses, they are not qualified to say anything about the spiritual. They are not qualified to say that evolution DID NOT happen by spiritual intervention. They resolve this question by sliding over from science into physicalism and asserting that the spiritual does not exist. This, of course, is not a valid argument.

There is always room in the universe for the spiritual, unless you know the whole universe, absolutely, in all its dimensions.

Eklal Kueshana has offered the explanation that angels created new species by genetic engineering when the environment was ready to support those new species.

This fits the evidence better than Darwin's theory, given that new species have appeared more quickly than can be explained by the slow process of adaptation. It also fits the evidence far better than the Biblical account of Creation. It acknowledges evolution, but says that the process was helped along by angels.

My friend asked me, "Have you ever seen an angel?"

There are three things wrong with this question. First of all, this is Eklal Kueshana's teaching, not mine. Presumably he has experienced angels. Second, I am presenting it here only as a working hypothesis, to be explored and tested. One is not required to supply proof before presenting a hypothesis. And third, it is probably not possible to see an angel because they exist at a higher vibrational rate than physical entities, according to Eklal Kueshana. But there are certainly indications that there are such things as angels.

I shall describe later my death experience, how I shot up into the presence of the Light and the figure of Christ. It certainly was very convincing that there were very real spiritual beings "up there," as I experienced it. The Catholic Church has defined nine levels of angels. New Age bookstores have whole shelves of books on the subject of angels. Some people I know have actually experienced angels. These people appear to be sane. So it is not logical, not valid, not scientific to simply dismiss the idea of angels.

Scientists seem to have been steered away from intelligent design by the word "supernatural," which carries with it the false premise that the "natural" is known. The "natural," to the physicalists, means the physical, and "supernatural" means anything non-physical, which to their way of thinking does not exist and is therefore imaginary.

Actually, everything that exists should be considered "natural." We don't know what that is, and it is science's job to find out. We know that Santa Claus, the Easter bunny, and the tooth fairy are imaginary characters invented by adults for children. But we don't know for certain about spiritual beings, such as ghosts and angels. These entities have only been dismissed by physical scientists, not proved to be imaginary.

Stephen Jay Gould dismissed Eklal Kueshana's explanation under the category of "Some personal versions of creation" (Gould, 1999, page 126). The word "personal" is used by authoritarian types as a put-down, to mean "not important," or "none of the rest of us care." *The Ultimate Frontier* by Eklal Kueshana had sold 200,000 copies by the time I became aware of it, in 1971. That's 200,000 more than "personal." The theory is certainly out there, and published, and publicized, and in major libraries, and far from "personal." In fact it reflects badly on Stephen Jay Gould as a scholar that he did not acknowledge this.

And then there is the trickery of labeling any argument in favor of creation "creationism" or "creationist," which is then defined by the weakest possible argument, usually the Fundamentalist interpretation of the Bible, that the earth and all living creatures were created by God in 6 days in 4004 BC, an argument that has already been disproved scientifically. Actually, so many invalid arguments have been advanced by religious people in favor of creation that the word "creationism" itself carries with it a stigma. Simply linking an argument to creationism by saying that creationists have used it is a device that is used to dismiss it. But if Eklal Kueshana's explanation is not dismissed, or ignored, or put down, or misrepresented, it can be seen that it fits the data much better than Darwin's theory.

All it is doing, really, is replacing the idea of "random mutation" with "genetic engineering." Well, actually, it is postulating angels, and a whole other dimension or dimensions in the universe. I guess that's just too much for some people to swallow.

But if there were such things as angels, operating on a whole higher level of existence than human beings, and if these angels had the ability to get into the substance of species and alter their genes, to know, first of all, precisely what alterations were necessary to create the desired species, then, yes, this process would take place faster and more predictably and allow new species to evolve in the time frame that has been observed.

I realize that Eklal Kueshana's theory is as radical in our day as Darwin's was in his day. And who would be qualified to test it? Biologists aren't qualified to test it. You would have to be able to read the Akashic Record, or something like that, to know whether angelic intervention occurred. Is there an Akashic Record? Biologists don't know.

And you would also have to be able to read the Akashic Record to be able to say with certainty that angelic intervention DID NOT occur. Unintelligent design is beyond the scope of biology either to prove or disprove.

So it appears that there are major problems with unintelligent design: First of all, the biologists may have disproved it already. Second, the use of the word "random" implies that they don't know the exact mechanism that generates new species, or whether it is different from the mechanism responsible for adaptation. And third, the use of the word "random" implies that they don't know whether that mechanism was a higher intelligence, and because biologists have limited their view to the known physical, they aren't qualified to determine that new species were NOT produced by a higher intelligence.

*   *   *

In my review on Amazon.com of *Why Evolution Is True*, by Jerry A. Coyne, my mention of "angels" was met with the most extreme ridicule I have ever witnessed in my life. The physical scientists have not only ignored the possibility that there might be such a thing as angels, but they have gone one step further and set up barriers of extreme social pecking, to prevent people from even asking the question. This ridicule might be justified if it was supported by solid scientific evidence, but it is not. It is simply an attempt to discredit the Target and destroy that person's credibility in the view of the Audience. In the academic community, this social pecking goes beyond ridicule. People can lose their jobs and their entire social status simply by asking the question, "Is there a higher intelligence in the universe, and might it have played some part in its creation?"

This extreme ridicule and the social degradation that goes with it create an extreme bias, where people are deterred from EVEN LOOKING FOR such a thing as angels. This bias works against impartial scientific inquiry. So the ridicule is more than just unscientific. Because of the bias it creates, it is actually anti-scientific.

Creation has to be ruled out by legitimate means, not by ridicule, bias, or force.

CHAPTER 6

Science as Religion

If unintelligent design does not explain how living species came to exist, then we don't know how that happened. Can't we just say, "I don't know?" No, we can't.

There are questions we don't know the answer to: How did the universe and all living creatures come to exist? Is there a God, and if there is a God, why is this God not supremely obvious to us every moment of our lives? What happens to us when we die?

I don't know the answers to these questions. Does anybody know? I don't know the answer to that question, either. Of course there are people that claim to know, but how would I know that they are right, unless I had that same level of knowledge?

To explain and deal with the unknown, we have religions, starting with primitive religions that have gods to explain all the forces of nature, especially the frightening ones, such as thunder and volcanoes and what makes the sun go down. And as we have understood the natural causes of these things, we have rid ourselves of these gods and moved to a more advanced God to explain deeper and more profound things, and the more subtle forces, such as the spiritual. This raises the question of whether the whole idea of "gods" or a "God" isn't just a way of explaining the unknown. When we really understand all the secrets of the universe, will we be able to do away with the concept of a "God" entirely? Or will we see God in all His Glory? We don't know that, either. Again, there may be people who know all that, but I, for one, don't know that they know.

Religion, where it represents the unknown, is primarily fiction. It has to be fiction, because this is the unknown. (When I say "fiction," I don't mean that it is false, but only that it is created out of people's imagination.) People have had glimpses of the spiritual, and upon these few glimpses whole cosmologies are built, which are claimed to be the absolute and ultimate truth. A large part of this has to be fabricated, again, because it is unknown.

The main characteristic of religion is that it is a rigid belief system. It has to be a rigid belief system, because if it ever changed, that would be an admission that it wasn't the absolute and ultimate truth in the first place. When I use the word "religion," I am talking about a rigid belief system.

The things we know about, we can argue about rationally, but the things we don't know, we have to fight about to the death.

I separate religion from the spiritual. Religion is the way we have dealt with the spiritual in the past. But having a rigid belief system isn't the only way to deal with the spiritual. It is possible to deal with the spiritual using the methods of science — to question, explore, experience, and learn. That way, we can reduce the area of the unknown, the same as we have done with the physical.

Stephen Jay Gould in *Rocks of Ages* says there is no conflict between science and religion, because they operate in "Non-Overlapping Magisteria." What he is saying is that science deals with the physical and religion deals with the spiritual. I disagree. Science and religion are really two different ways of thinking. Religion is a rigid belief system, whereas science is open to new discoveries, new knowledge, and possibly overturning old beliefs. There will always be a conflict between a rigid belief system and a system such as science which is always growing and expanding our knowledge.

My knowledge of religion is limited mostly to the Christian religions and the Judeo-Christian Bible. If other religions don't operate the same way, you can add the word "Christian" every time I say "religion." But in the context of physicalism and Darwinism, "religion" means the Christian religions, because physicalism and Darwinism grew out of Western civilization, which has been dominated by the Christian religions.

If religion is primarily fiction, then why do people believe this fiction?

I guess the first answer is that they have been indoctrinated with it as small children. It took me more than 50 years of breaking away from the grip of religion even to see clearly enough to ask the question.

And then the question becomes, how are people indoctrinated?

And the answer to both questions is the same:

It all goes back to when we were very small children and our parents were like gods to us, all-powerful and all-knowing. We looked to them for all the answers. The idea that it was possible for somebody to have all the answers was imprinted on our minds. We carry with us a residue of that time. Then, as I have said before, this is reinforced when we go to school, where the teacher has all the answers, and our place is only to memorize. Sunday school is the same, but now we are being told to believe on an even higher authority, the Word of God. So we have all been conditioned to believe that this kind of authority can really exist.

We believe religion as an explanation of the unknown, because we don't know, and somebody seems to know, with great certainty. So we believe on the basis of their claimed authority. There is a social need for this kind of authority in dealing with the unknown. It makes people feel more comfortable and secure to know that somebody has all the answers. If all of the priests and preachers suddenly disappeared one day, there would be a social need to replace them.

This social demand for "authorities" is at the root of many social problems, including religious intolerance, war, and dirty science. It isn't so much the fault of the scientists who play the role of "authorities" as it is the fault of the people who believe them. With the kind of conditioning we have had since early childhood, it is not surprising that people believe there are "authorities" who have all the answers.

But at some point in our growing-up process it would be desirable for us to recognize that these teachers and preachers are human beings just as we are, making human errors just as we do, and that our all-knowing parents were not really all-knowing, but only appeared to be, relative to what we knew as very small children.

That revelation came to me in psychotherapy. I was furious at my boss, because he had made some kind of mistake. I realized I was angry because I saw him as the idealized parental figure from early childhood, expected to be perfect. When I was able to see him as a human being like myself, I forgave him for his mistakes, and the anger went away.

In the same way, scientists are human beings like the rest of us and subject to errors and bias and prejudice and in-group pressures and psychological needs to dominate. It may take psychotherapy for people to recognize that scientists are not idealized parental figures speaking Truth with authority.

It may take psychotherapy even for people to recognize what is authority and what is not. Science is an authoritarian system. Scientists prove things to be believed by other people, as differentiated from people proving things to themselves. We can all use scientific methods to prove things to ourselves. First-hand knowledge is our primary source of knowledge. But science provides us with an important secondary source of knowledge. We can believe what scientists have proved ("verified") by the scientific method, with the same reservations as we would have about things we have proved to ourselves: We are all human beings and subject to error, even though science, as a system, tends to minimize errors. There is legitimate authority in what scientists have proved ("verified") by the scientific method, although not the 100% perfection we would expect and demand from idealized parental figures.

Outside of what has been determined by the scientific method, scientists have no legitimate authority at all. Outside their fields of expertise and without reference to scientific studies or findings, their opinions are no more valuable than those of the man in the street.

There are two meanings of the word "authority." One refers to knowledge and the other refers to power. In early childhood, when our parents were as gods to us, they had both. They were both all-powerful and all-knowing. And as small children, we may not have even differentiated these two attributes. But as we grow up, I think most of us learn the difference.

Or maybe we don't learn the difference, because the beliefs of religions and in-groups are enforced by social pecking. Historically, heretics and psychics were burned alive, the most painful kind of death, as I have read. If you look up "torture" on the Internet, you will find other horrible things that people did to other people, simply because those who were supposed to believe did not agree with those in power. So knowledge, historically, has been controlled by power. The Truth has been what those in power have said it was, enforced by unbearable tortures. It is only recently that governments have granted to individuals the freedom to speak and write what they believe. And that system doesn't quite yet work, because persons recognized as "authorities" can severely damage other people's reputations, and most of us don't have enough money to sue them.

I would like to use the word "authoritarianism" to mean belief on the basis of perceived "authority," but the word "authoritarianism" is defined in my dictionary (Merriam-Webster, 1996) strictly in terms of power, the power of governments that have the very real authority of life and death over individuals. We have no word to mean the fake authority of individuals and religions that have set themselves up as "authorities" but really don't know what they are talking about.

But we have freedom of religion, so that if we don't like what our religion is telling us to believe, we can move to another religion, and not worry about being burned alive. Even in this case, the word "religion" is restricting. Does this mean "religion" as I have defined it, and as it has historically existed, as a rigid belief system, claiming to be the absolute and ultimate truth, and enforced on its members by extreme pain or threats of extreme pain such as burning in Hell forever? Or can we be "loose constructionists" and extend the legal meaning of "religion" in our Constitution to include any belief system? Only with freedom to have any beliefs and speak or write those beliefs without social pecking can we separate knowledge from power and extend the meaning of the word "authoritarianism" to mean "belief on the basis of perceived authority."

So it should be easy to see, given this psychological need of human beings to replace idealized parental figures with authority figures, that when science swept away the old religious mythology, something had to replace the old religious "authority." Science became the obvious choice, and scientists were swept into this social role as priesthood. They were a perfect fit for the job: They could claim that nothing existed beyond the physical, and their knowledge of the physical was certainly authoritative. But to the extent that they assumed this role, they ceased to be scientists and became physicalists.

I suggest an alternative that would be more appropriate for people who would like to preserve their integrity as scientists. I have adopted the cosmology set forth by Eklal Kueshana, known as "the Brotherhoods' philosophy." The cosmology itself is not important. (It is a more plausible version of the Christian mythology.) What is important is that the first rule of this cosmology is that it is to be treated as a working hypothesis, and not as absolute truth which has to be believed absolutely.

So I make the suggestion to physical scientists that they treat their belief in a purely physical universe as a working hypothesis only, so that they can be open, as true scientists, to evidence which might contradict it.

There might also be the thought among physical scientists that to concede that there are spiritual aspects to reality means going back to the old religious beliefs. That doesn't have to happen. The culture is moving forward.

Carl Jung managed to break through the domination of physicalism to discover spiritual aspects of human beings, independent of religion. Generations from the Hippies onward have recognized that religion and the spiritual are two different things. Studies are being carried out in subjects such as reincarnation, the power of prayer, communication with spirit entities, and spiritual healing, without making any reference to the Bible or postulating the existence of God. True scientists should be welcoming these studies as they help us break through the dogma of a purely physical universe and encompass a broader view.

CHAPTER 7

The Mental Senses

Physicalism has steered psychology away from the study of the mind by insisting that the only legitimate evidence is the evidence of the physical senses. Thus limited to physical evidence alone, psychology has ceased to be the study of the mind and has become a mess. I have spent years trying to untangle that mess, and I need to take a few chapters here to explain the mess and how it can be straightened out.

* * *

How do I know that I have a mind? How does anybody know that they have a mind? The only way we know that we have such a thing as a "mind" is by being aware that we have certain mental processes — our thoughts, our emotions, our memories, and our dreams. Whatever is performing these processes is what we call "the mind." To be able to be aware of these thoughts, feelings, memories, and dreams, we must have some mechanisms of perception. Just as we have "physical senses" to perceive the physical world, I define "mental senses" as those mechanisms by which we perceive our mental processes.

I don't think that the culture is fully aware that we have senses to observe the mind, because we have the expression "the five senses" to mean the five physical senses, and then the expression "sixth sense" to mean something psychic, something abnormal. There is no room in between, in the language, for those mental senses that every normal person has.

Descartes said, "I think; therefore I am," ("Cogito ergo sum"), because he distrusted the evidence of the senses. But how did he know he was thinking, unless he first sensed it?

The early psychologists studied the mind with what they called "introspection," which has a very convoluted definition, which includes thought and imagination, as well as perception. They used the mental senses, but they weren't totally clear in thinking about what these mechanisms were, and they didn't clearly separate the function of perception out of the convoluted definition of "introspection." If they had been more fully aware of the mental senses, they would have had a better argument with which to resist the domination of physicalism.

The discovery of the mental senses is my own independent discovery, one of many in my efforts to turn my back on the culture and design a new civilization from scratch. I recognized that in determining the truth of a proposition such as "If $\underline{A}$ is greater than $\underline{B}$, and $\underline{B}$ is greater than $\underline{C}$, then $\underline{A}$ is greater than $\underline{C}$," I don't compare the size of the letters on the page, but construct some kind of images in my mind and then compare those images. The ability to "look at" those mental images is what I call a "mental sense." Similarly, when somebody explains something to me and I say, "I see," I am seeing something mental and not physical.

This ability to "see" something mental is something separate from the ability to construct a mental image or the attention required to do this or the volition required to want to do this. I believe that Western philosophers have created unnecessary confusion by bundling all these things. For example, I have read that observation requires attention, and that attention distracts from the perception. These mental perceptions are an involuntary thing, just as my physical senses are perceiving the external world, whether or not I am paying attention to it, and whether or not I want to be perceiving the external world at this moment.

Locke and Kant had views of mental senses similar to my own:

Etymologically, the term "introspection" — from the Latin "looking into" — suggests a perceptual or quasi-perceptual process. Locke writes that we have a faculty of "Perception of the Operation of our own Mind" which, "though it be not Sense, as having nothing to do with external Objects; yet it is very like it, and might properly enough be call'd internal

Sense" (1690/1975, 105, italics suppressed). Kant (1781/1997) says we have an "inner sense" by which we learn about mental aspects of ourselves that is in important ways parallel to the "outer sense" by which we learn about outer objects. (Schwitzgebel, 2016, section 2.2)

But then these perceptions that Locke and Kant identified somehow exploded into "introspection," which is defined by Merriam-Webster as synonymous with "self-contemplation, self-examination, self-observation, self-questioning, self-reflection, self-scrutiny, self-searching, soul-searching" (Merriam-Webster, 2017). From there, it is my opinion that Western philosophers have created mostly confusion on the subject.

*   *   *

Science requires, first of all, observation of the object being studied. To study the mind scientifically, one must first observe the mind, and that is done with the mental senses. I don't see any other way of observing the mind. So, until we find a better way, any scientific study of the mind must start with the mental senses.

I have heard that Western philosophers have questioned the accuracy of the mental senses. First of all, if they haven't accurately pinpointed the mental senses, then how do they even know what it is that isn't accurate? Accurate or not accurate, the mental senses are what we have with which to observe our minds.

Western philosophers have also pointed out inaccuracies in the physical senses. We see mirages. Some people are color-blind. Some people are tone-deaf. We don't see the ultimate reality of things, but only their surface appearances. In spite of these inaccuracies, physical science manages to operate with the physical senses. In the same way, mental science can produce accurate results with the mental senses, by recognizing inaccuracies and avoiding them.

Present-day psychologists are wrong if they think that observing brain activity with their expensive equipment is the same as studying the mind, because their expensive equipment does not see thoughts, feelings, memories, or dreams. The person under observation has to do the observing, with his/her mental senses, of what he/she is thinking, feeling, remembering, or dreaming, and then communicate that to the scientists, in order for them to know what is happening mentally while they are observing the brain activity. Maybe in some science-fiction future somebody will invent a device that will record brain activity and play it back as mental processes, but for now we are very much dependent on the mental senses to observe mental processes as people know them. And even when we can play back such a recording, the "subject" will have to observe the playback with his/her mental senses and say "Yes, that is what I was thinking / feeling / remembering / dreaming," in order to validate it.

But the physicalists have been trying to convince us that only observations with the physical senses are legitimate for scientific study. They have done this through assertions, bad logic, and manipulative tactics, leaving many gaps in the paper trail.

First of all, there is the obvious limitation of the mental senses: The scientist can observe at most his/her own mental processes, unless he/she is telepathic. So, to observe many people's mental processes, the early psychologists had to train their subjects to be observers of their own mental processes. The first problem with this was that the psychologists trained them in different methods of observation, so that results couldn't be replicated by other psychologists. Second, the subjects made errors in their observations. And third, some subjects lied about what they were observing.

CHAPTER 8

John B. Watson, Scientist/Manipulator

John B. Watson came to the rescue of psychology in 1913 with his famous paper, "Psychology as the behaviorist views it," saying that psychology should become the study of physical human behavior observable with the physical senses. His reasoning made perfect sense:

> ... What we need to do is to start work upon psychology, making *behavior*, not *consciousness*, the objective point of our attack. Certainly there are enough problems in the control of behavior to keep us all working many lifetimes without ever allowing us time to think of consciousness *an sich*. ... (Watson, 1913, pages 175-6)

Please note that he is not saying here that introspectionism is invalid, much less presenting any argument to demonstrate that introspectionism is invalid. He is only saying, modestly, that we should see what we could learn by studying human behavior. He gives no reason why the study of the mind with the mental senses should be abandoned. And yet people abandoned the study of the mind with the mental senses. I have to speculate here because I can't find any document giving logical reasons, but it seems that Watson's influence was a major factor.

By 1925, when Watson published *Behaviorism*, he had simply determined that there was no such thing as the mental:

> ... there is the misconception going the rounds that there is such a thing as the mental ... (Watson, 1925, page 243)

He slides this assertion as a subordinate clause into a larger assertion dismissing the idea of mental illness. Here is the full context:

> As long as there is the misconception going the rounds that there is such a thing as the mental, I suppose there will be mental diseases, mental symptoms and mental cures. ... (Watson, 1925, page 243)

His sole argument for the nonexistence of the mind is because, he says,

> The behaviorist advances the view that *what the psychologists have hitherto called thought is in short nothing but talking to ourselves.* ... (Watson, 1925, page 191)

He claimed that thinking was only subvocal talking.

This is a huge assertion. He is sweeping away the mind and all its accomplishments in one sentence.

This assertion grabbed me for a long time before I was able to break out of it. I believe that John B. Watson was one of those people with strong mentalities and hypnotic powers who are able to block other people's thinking and implant hypnotic suggestions in their heads. I believe that the many unsupported assertions he left in the minds of academic people are a testimony to that.

I was able to break free of Watson's assertion first of all because it implies that we think only in words. It is very difficult for me to put my thinking into words, and it makes me furious when anybody claims that we think only in words.

And then there was the experience of explaining to a friend how to know that the two sections of the extension ladder were successfully locked into place. I was gesturing with my hands, putting them into the same plane, and using the words, "plane," "equal," and "parallel." Finally, right now, years later, I am composing the words, "The extension ladder is successfully locked into place when the surfaces of the steps in one section of the ladder are in the same plane as the surfaces of the steps in the other section of the ladder." I certainly didn't have those words when I was gesturing with my hands, although I knew exactly how the process was supposed to work.

Watson is tricky here in his use of the word "thought," because not all mental activity is thought. I can remember a face, or experience a dream, or feel an emotion. So his logic has serious leaks in it.

One of the reasons we dream in symbols is because the dream is trying to tell us something for which the culture has no words.

An associate in the computer field once asked me how I designed computer software, and my answer was, "Primitive art." I go deep in my head and see pictures and patterns. These then translate into flow diagrams, and from the flow diagrams I write the computer code. The words definitely come last.

> ... How do we ever get new verbal creations such as a poem or a brilliant essay? *The answer is that we get them by manipulating words*, shifting them about until a new pattern is hit upon. ... (Watson, 1925, page 198)

This sounds like the attempts people have made to get computers to write poetry by combining random words in likely patterns — with interesting but bizarre and meaningless results. A real poem comes from the depths, and all but the most gifted grope for words to express the feelings.

Did I say "feelings?" Where are feelings in Watson's theory?

If the mind doesn't exist, then there is no such thing as scientific thinking; there is only scientific talking.

I have seen evidence that many of Watson's ideas are still implanted in the minds of academic people, with no evidence to back them up:

1. We think only in words.

2. There is no such thing as creativity. It is all just a synthesis of known elements. (If we only think, and not create, and if we think only in words, then creativity has to be just a recombination or rearrangement of words, as Watson stated in the quote above.)

3. There is no such thing as intuition. (If we think only in words, then we don't think in pictures or patterns.)

4. Self-education is impossible — a delusion, a myth. (If we think only in words, and words of course come from the culture, then there is no thinking beyond or outside of or independently of the culture.)

I think some academic people need to examine the foundations of some of their beliefs.

*   *   *

I got the idea that Watson might be a manipulator when I read the following in a psychology textbook:

> Watson argued that introspection was, if anything, the province of theology. (Alloy et al, 1999, page 123)

This led me to check out Watson's major works. As I have already said, his 1913 paper, "Psychology as the behaviorist views it," I thought was very reasonable.

His 1919 book, *Psychology from the Standpoint of a Behaviorist*, is very technical. I don't see much to complain about, except on page 396, where he refers to "the Freudian mystics." and on page 1, where he is trying to make all of psychology into behaviorism:

> Psychology a Science of Behavior. — Psychology is that division of natural science which takes human activity and conduct as its subject matter. ... (Watson, 1919, page 1)

But in *Behaviorism* he starts becoming very free with his logic and his theories, and in my view very manipulative with his audience. Here is where he makes the connection between psychology and religion:

> ... Indeed we should point out at once that behaviorism has not as yet by any means replaced the older psychology — called *introspective psychology* — of James, Wundt, Kulpe, Titchener, Angell, Judd, and McDougall. Possibly the easiest way to bring out the contrast between the old psychology and the new is to say that all schools of psychology except that of behaviorism claim that *"consciousness" is the subject matter of psychology*. Behaviorism, on the contrary, holds that the subject matter of human psychology is the *behavior or activities of the human being*. Behaviorism claims that "consciousness" is neither a definable nor a usable concept; that it is merely another word for the "soul" of more ancient times. The old psychology is thus dominated by a kind of subtle religious philosophy. (Watson, 1925, page 3)

So here, with a hop, skip, and jump, he goes from "consciousness" to "soul" to "religious." That is his complete argument. Thus he misrepresents the introspective psychology in a destructive way: Using the word "religious" to describe a branch of science is simply a smear tactic. And this inaccuracy has survived long enough to be reported in a psychology textbook 75 years later.

Then, in a section called "The Religious Background of Current Introspective Psychology," he goes on to ridicule religion, saying how certain lazy individuals were able to exploit people's fears and become medicine men, and thus not have to work. He concludes his satire on religion with the following:

> ... It was the boast of Wundt's students, in 1879, when the first psychological laboratory was established, that psychology had at last become a science without a soul. For fifty years we have kept this pseudo-science, exactly as Wundt laid it down. All that Wundt and his students really accomplished was to substitute for the word "soul" the word "consciousness." (Watson, 1925, page 5)

Watson quickly flips here from "science" to "pseudo-science" (the smear word) without taking the trouble to explain how the transformation was made. He then repeats the assertion that "consciousness" is "soul," maligning Wundt and his students in the process, and perhaps exploiting the doubts of his own students. So who are you going to believe? Who is giving the grade in this course?

Contrast the "pseudo-science" with what recent textbooks say about Wundt:

> Wilhelm Wundt, the founder of experimental psychology ... (Lazerson, 1975, page 337)

> In short, with Wundt, psychology had evolved from the *philosophy* of mental processes to the *science* of mental processes. (Bernstein et al, 1997, page 7)

Many of the early psychology books in my great-grandfather's library did go on excessively about the "soul," but to slide this attribute over to people such as Wundt, Kulpe ("systematic experimental introspection"), Titchener, and McDougall is a misrepresentation.

Watson then starts attacking William James, something he does throughout this book.

Note that Watson's field is behaviorism, but he is talking here about introspectionism, the competition, to make its adherents look bad. If he was using valid arguments, that would be one thing. But he is just making assertions to discredit the opposition. This is an example of what I call "mental warfare." And I can't see that he needs to discredit the opposition in order to teach behaviorism. His rationale was perfectly spelled out in his 1913 paper — let's just see what we can learn by studying human behavior. If the study of human behavior was so superior, then it would have proved itself, without any need for trashing the opposition.

On page 200, he brings in the name of "Mrs. Eddy" to make a point, knowing, I'm sure, that Mary Baker Eddy is somebody his students have already been conditioned to reject.

In the early edition of *Behaviorism*, he links Freudian "psychoanalysis" to Voodooism (Watson, 1925, page 18). This was removed in later editions, I suppose because this was a stretch even for Watson.

After pages and pages of what I see as a propaganda campaign against the "mental" and "consciousness" and "introspection," Watson makes the following statement:

> ... The unscientific nature of Freud's conception is surely apparent to you ere this. If it is not I'll have to give you up. (Watson, 1925, page 242)

This is the kind of manipulative expression that adults used to use with young children before people learned that it could be psychologically damaging (i.e. "abandon you").

He goes on:

> ... I venture to predict that 20 years from now an analyst using Freudian concepts and Freudian terminology will be placed upon the same plane as a phrenologist. ... (Watson, 1925, page 243)

Actually, 35 years from then I was enjoying good results, life-changing results, with a psychiatrist using Freudian concepts and Freudian terminology — updated of course, as the 1960 automobile was an improvement over the 1925 automobile. And Freudian psychology is still alive and well and respected, despite all the bashing. Karmic justice is served here, as Watson falls victim to his own propaganda.

There are probably many like him who take advantage of their position of power to manipulate innocent young undergraduates, but they aren't all so foolish as to put their manipulations in print.

CHAPTER 9

The Pretense of Objectivity

Somehow the idea has been implanted in the minds of scientists that the observation of physical phenomena with the physical senses is "objective" and the study of mental phenomena with the mental senses is "subjective." The rationale for this is that mental phenomena can be observed by only one observer, whereas physical phenomena can be observed by multiple observers. That makes the observation independent of any one observer and therefore "objective." The flaw in the argument is in calling it "objective."

The word "objective" refers to the object being studied. The word "subjective" refers to the subject doing the study. The word "objective" means the characteristics of what the object really is, as opposed to "subjective," which refers to how the subject sees it. By applying the word "objective" to the perception of the object, people are implying that the object is being perceived with 100% accuracy, just because multiple people can be perceiving it. This is a logical and mathematical error.

Scientists have maintained the appearance of "objectivity" by speaking and writing in the impersonal mode, as in "It has been determined," instead of "I have determined." Rupert Sheldrake, in his chapter "Illusions of Objectivity," says that children in England are being taught to write in the passive voice, as in "The test tube was heated and carefully smelt" (Sheldrake, 2012, pages 299-301). But there is always an "I" doing the observing, and always an "I" drawing the conclusions from those observations.

All perceptions are subjective. All perceptions are made by the subject. Some perceptions are made by instruments, which might be said to be "objective," in the sense that they are free from bias. Yes, but ultimately subjective human beings have to read those instruments and interpret those readings. And in the beginning those instruments were designed by human beings. In any perception of reality, the human element is always present, and that human element is always subjective and always carries with it the possibility of bias.

The presence of multiple observers does not reduce the probable margin of error to zero, but reduces it by a factor of the square root of the number of independent observers. What we have is not an "objective" perception of reality, but a "collective subjective" perception. The probable margin of error is reduced, but not eliminated.

A great deal of the bias in how we see objective reality is simply in how human beings see things. As physics dives deeper into the physical nature of reality, it becomes more and more obvious that our perceptions of reality are really a gross approximation of what reality really is — enough to enable our species to survive (so far), but that's the best that can be said for our perceptions. So all observers share this human bias, which does not average out.

And did I say "independent" observers? Scientists are all part of an in-group, who all share the same biases, as I have already explained. They are far from being independent observers, and these shared biases also do not average out.

So this collective subjective perception is really a long way from an objective perception, implying absolute truth. "Objectivity" is a pretense.

In theory, yes, multiple observers can observe physical phenomena, but in reality they usually don't. Most scientific observations are made by one person. The way science achieves consensus is through replication. Other scientists make the same kinds of observations, and they see if they all agree.

Scientists don't all study the same gorilla. They each study their own gorilla and then compare notes.

So if all observations are subjective, and if most scientific observations are made by one person, then observations with the mental senses fit into that same category of scientific possibility.

By replication, the self-observations of multiple experimental subjects can be compared to see where they agree.

Yes, I can lie to you about what I experienced in my own mind. But also scientists can lie to you about their observations of gorillas in a remote jungle. The great thing about science is that replication gets rid of things such as intentional lies and fraud the same way it gets rid of innocent errors, simply because the inaccuracies are not repeated in other people's observations.

I define "mental science" as the study of mental phenomena with the mental senses, as contrasted to "physical science," which is the study of physical phenomena with the physical senses. Mental science can be accurate, just as physical science is accurate, by applying the scientific steps of observation, accurate logic, and replication. It is only the pretense of objectivity that makes physical science appear infinitely more accurate than mental science.

CHAPTER 10

How Psychology Abandoned the Study of the Mind

The first step in doing science is observation of the object being studied. Even if "the mind is nothing but the physical brain," as the physicalists assert, the mind and the brain are not the same thing.

To illustrate what I mean, think of a violin. The violin is like the brain, and the mental processes that we know as "the mind" are like the music produced by the violin. This is not a perfect analogy, but at least it serves to separate the two.

The music is not observed by looking at, measuring, and analyzing the violin or even looking at, measuring, and analyzing the vibrations created by the violin. The music is observed only by listening to it. The vibrations can be translated into music, as in a recording, but that music is only "music" when it is being listened to by the human ear.

In the same way, the mind is only the mind when it is being experienced as mental processes, such as thoughts, emotions, memories, and dreams, and these mental processes are observed only through the mental senses. If we can at some future time play back a brain scan of a memory or a dream so that it can be shown on a screen, with all its sights, sounds, colors, smells, sensations, and emotions, then we will have bypassed the mental senses. But in order to reach that state, considerable use of the mental senses will have to be made in order to make sure that that recording is accurate.

*   *   *

How did psychology go from the study of mental processes to the study of the physical instrument? And why was the study of mental processes abandoned? I have not seen the answer spelled out anywhere, as "Introspection was abandoned for reasons x, y, and z." So I am providing the definitive explanation here, out of my creative imagination.

Historian of psychology Duane P. Schultz wrote in 1969:

> ... No psychologist today calls himself a behaviorist — it is no longer necessary to do so. To the extent that American experimental psychology is today objective, empirical, reductionistic, and (to some degree) environmentalistic, the spirit, if not the letter, of Watsonian behaviorism lives on. ... (Schultz, 1969, page 236)

But this doesn't explain why they all abandoned introspection and became behaviorists, or why present-day mainstream psychology is devoted to brain science. And I can't be sure that there aren't some psychologists somewhere still practicing introspectionism, although it would seem that they would have to be unemployed, unfunded, and unpublished. What I am sure of, though, is what I gathered from reading "Criticisms of Psychodynamic Theory" in a 1999 textbook:

> ... most of its claims have never been tested in scientifically controlled experiments.

> ... psychodynamic theory is not altogether closed to empirical testing ... (Alloy et al, 1999, page 107)

Although the author doesn't actually say "physical evidence," I inferred from the words "tested in scientifically controlled experiments" and "empirical testing" that physical evidence was what was meant, because that is the normal usage of those terms within the scientific in-group.

The rules of evidence have changed, so that physical evidence is the only evidence recognized for scientific work. The word "physical" doesn't have to be stated. It goes without saying that evidence must be physical.

An academic person once told me, "There is no referent," when I was discussing mental phenomena. This may have been the same attitude that John B. Watson assumed when he declared that the mental did not exist. If one is dedicated, as a scientist, to purely physical evidence, then it may help to simply deny that anything exists that cannot be perceived with the physical senses. I have also experienced this with a scientist who declared, "Dreams don't exist." This denial, of course, is physicalism at its extreme. Where it may serve to focus one's attention on the physical, it blocks people from a larger view of reality.

I don't think that Watson's dogmatic assertion that the mental did not exist was enough to cause psychologists to change their rules of evidence, but I do think that the certainty that they were doing established, approved science was an important factor in their switch from the mental to the physical.

First of all, there was some debate over whether introspection was a legitimate source of scientific evidence. There was no clear recognition of the mental senses. The physical senses were the only senses known to exist for scientific observation. In the consciousness of the times, there was only the assertion that observation with the physical senses was "objective," whereas introspection was "subjective." I am sure that this assertion steered many psychologists, including Watson, away from introspectionism, because they just felt that they were being more "scientific" by observing physical phenomena with the physical senses.

Also, in the 1950s, when I first became aware of psychology, physical scientists were ridiculing psychologists, saying that they weren't doing science, because their knowledge wasn't based on physical evidence. And the psychologists, instead of meeting that ridicule with a strong argument in favor of mental evidence, subordinated themselves to that social pressure, calling their introspective observations "theories." Thus they put themselves in a very weak position, where the physical scientists could claim that these "theories" had no scientific basis, and that psychologists had to conform to the rules of physical evidence if they were to call themselves "scientists" at all.

But let's look at science again. The first rule of science is evidence, and that evidence is acquired by observation of the subject matter to be studied. The observation is appropriate to the subject matter, and for each subject, the methods of observation are different. Biology requires microscopes to study organisms at the cellular level. Chemistry is based on a Periodic Table which came out of somebody's creative imagination, but which seems to work every time. Physics is based heavily on logical and mathematical inferences. Nobody actually observes the atomic and subatomic entities they are working with.

So each field of scientific study has its own kind of evidence, appropriate to the subject matter being studied. The rules of evidence are determined by people expert in the field. Similarly, the scientific study of the mind has its own particular kind of evidence, which should be determined by people who are expert in that particular field. Physicists and biologists are not qualified to tell people how to study the mind. In particular, the scientific study of the mind requires first of all observation by the mental senses. This scientific study should not have to submit to the domination of physical scientists asserting that all evidence, in order to be called "scientific," must be evidence from the physical senses. The study of the mind, as observed by the mental senses, should also be called "scientific," as long as every effort is made to make accurate observations and interpret them with accurate logic.

In the absence of any paper trail, then, this is my fictional explanation of why psychologists abandoned the study of the mind with the mental senses:

Psychologists have never been fully aware that there are such things as mental senses, that this is the only way to observe mental processes, and that doing so is scientifically valid.

While they were debating whether introspection was a scientifically credible means of observation, they were unable to formulate solid rules of evidence for the study of the mind.

John B. Watson reinforced their knowledge that observing physical phenomena with the physical senses was the established and accepted way to do science.

Ridicule from physical scientists telling psychologists that they weren't really doing "science" put social pressure on them to conform.

The constant propaganda campaign, asserting that observation with the physical senses was "objective" and therefore accurate, whereas introspection was "subjective" and therefore biased, had a hypnotic effect because of its constant repetition. Because psychologists were unsure about introspection in the first place, their critical faculties were not able to reject the pretense of "objectivity" and the air of certainty with which the assertions were repeated. When they began to call their observations of mental processes "theories," it was the beginning of the end of psychology as the scientific study of the mind.

With this constant bombardment of certainty against uncertainty, psychologists were forced to concede that physical evidence was the only legitimate scientific evidence.

In present-day brain science, there is a strange hypocrisy: Psychologists on the surface are studying physical activity with the physical senses, while at the same time, in order to relate any of that to mental activity, they have to rely on the mental senses, asking their experimental subjects, "What were you thinking, feeling, dreaming, remembering?" The mental lives on, somewhere in the subconscious. Psychology needs to be psychoanalyzed.

CHAPTER 11

Freud and Jung
and Psychotherapy

While the psychologists were having problems with introspection and rejecting it in favor of something more "scientific," Sigmund Freud and his followers were learning something from mental evidence, and even learning something about the nature of its inaccuracies.

The mental senses were used successfully by Freud in developing a huge new body of knowledge, namely:

1. People bury their unwanted thoughts, feelings, desires, and motivations in the subconscious.

2. These repressed aspects of their personality continue to operate from the subconscious, in conflict with their conscious desires and motives.

3. These conflicts are caused by unresolved childhood traumatic experiences.

4. Dreams are not just meaningless nonsense. Dreams are a clue to these subconscious elements, "the royal road to a knowledge of the unconscious," as he put it.

5. People employ "defense mechanisms" to avoid seeing their subconscious motivations and their true selves. Anna Freud identified a dozen or so of these "defense mechanisms," including "repression," "projection," "displacement," "rationalization," "sublimation," and "denial," described in any psychology textbook (Alloy et al, 1999, page 96).

6. Freud developed the method of "psychoanalysis," using dreams and free association, in a confidential and non-judgmental

setting, to get rid of these conflicts by slowly and patiently bringing them out into conscious awareness and dealing with them there.

Carl Jung, Freud's number one pupil, carried Freud's tradition further and corrected many of Freud's errors: No, psychological problems don't all have to do with sex. No, dreams are not "wish fulfillment," but rather a correction to, or compensation for, the conscious attitude. No, the subconscious is not just a place where we bury our unwanted thoughts and feelings, but also a source of ideas, inspiration, and spiritual knowledge. When I first read Jung, I thought he had not yet broken free from the religious beliefs of the pre-scientific era, but actually he was moving ahead into a new era, an era where people can believe in the spiritual not on the basis of "faith" or somebody's authority, but on the basis of EVIDENCE. His discovery of "original experience," whereby one is able to see God for oneself, is potentially his most valuable (but not yet recognized) contribution to the culture.

I know that these things are true because I have experienced them myself. I have successfully completed psychotherapy, first to the satisfaction of my psychiatrist, and then to my own satisfaction, using the disciplines I had learned in psychotherapy and inspired by the writings of Carl Jung, analyzing my own dreams and doing journaling, until I was totally comfortable with myself as a human being. I have been through a process roughly equivalent to Jungian analysis.

If the evidence of the mental senses were to be accepted, it would be seen that I have replicated many of the discoveries of Freud and Jung in my own experience. This isn't "theory;" this is evidence. I have observed my own personal psychology directly with my mental senses. Unlike psychotherapists who must infer their clients' mental processes from the outside, my view of psychology is from the inside.

I have been advised by an experienced science editor not to include my personal experiences in this book, and especially not the fact that I have been through psychotherapy. But this is the only real evidence I have, to refute those people who are trying to dismiss Freud and Jung, saying there is no such thing as the subconscious, that childhood traumatic experiences don't determine our adult lives, that psychotherapy is only conversation, and that nobody ever benefited from Freudian methods. These opinions have been advanced, in all cases, by people who show no evidence of having successfully completed psychotherapy, and in most cases show evidence of needing it. I think the testimony of people who have successfully completed psychotherapy is needed, to counter this flow of misinformation. Would you rather learn how to climb Mount Everest from people who have actually climbed it, or from people who failed, or never tried, or claim Mount Everest doesn't exist?

Here are some of the discoveries of Freud and Jung that I replicated, with some variations, in the course of my psychological development:

Yes, there is a subconscious. But no, the subconscious is not a fixed place in the mind. I have observed thoughts gradually becoming conscious, over time, like fish emerging from a deep pool. As one develops psychologically, more and more becomes conscious. In therapy, I uncovered from my subconscious mental attributes I didn't know I had — creativity and intuition and will.

Yes, I have experienced the subconscious, from Freud's viewpoint, as a garbage dump where unwanted thoughts are buried, and I have also experienced the subconscious, from Jung's viewpoint, as a creative source of material that I didn't put there, including spiritual insights.

Yes, traumatic experiences in childhood shape our adult lives. I had not one but several, ranging from mild to crippling.

Yes, I lay on a couch and talked to the ceiling, with the psychiatrist behind me. It was more comfortable than sitting facing somebody. When I am trying to reach for deep thoughts, it is always distracting to be looking somebody in the eye. I can actually be more honest if I look away (contrary to folk belief).

My psychiatrist trained me as an observer of my own mental processes — thoughts, feelings, fantasies, dreams, and defenses. He trained me to speak in specific language, not abstractions.

My psychiatrist and I engaged in a constant dialog, sometimes an argument, as opposed to some methods of treatment in which the psychiatrist hardly says anything.

Freud's view of dreams as "the royal road to the unconscious" is accurate, but his view of dreams as "wish fulfillment" is not.

Jung's view of dreams as "a correction to, or compensation for, the conscious attitude" proved to be more accurate, and led me to a whole education from my dreams.

I never experienced "infantile sexuality." Sexuality for me started at puberty.

I did experience the Oedipal triangle, though, which has nothing to do with sex, but is the child wanting exclusive possession of his mother.

I know that there are thousands or millions of people who have successfully completed psychotherapy. If more of you would come forward and report your observations publicly, we would have a formidable body of knowledge to support the findings of Freud and Jung. But I know there are obstacles:

First of all, there is the arbitrary ruling by the scientific establishment that "anecdotal evidence" is not allowed. They dismiss it by saying it is "useless." That is the complete argument. But all evidence is evidence. Each individual experience represents a sample size of one. By itself, it is not significant, but as these individual experiences accumulate, they become a sample size that can be held to be representative of a population.

There is also the question of the credibility of the witness. If the town drunk reports having seen a sea serpent, we would view it with some skepticism, whereas if a person known to be reliable claims to have seen the same sea serpent, we would take it more seriously. So we might assign a coefficient of reliability to each report, depending on the individual and whether or not the observation is within the individual's area of knowledge. By arbitrarily dismissing "anecdotal evidence," scientists are effectively assigning a reliability coefficient of zero to every such observation. This is simply bad math. This is creating as much of a bias in one direction as believing every Bigfoot report would in the other. This is just another example of scientists blanking out potential knowledge.

Members of the scientific in-group complain that out-group members are not trained observers. But my psychiatrist trained me as an observer of my own mental processes. Just the fact that I succeeded at psychotherapy would indicate that my observations were accurate, at least accurate enough to have gotten past the primary source of error, my own psychological defenses. The same should be true of all people who have successfully completed psychotherapy. They would be excellent candidates for this kind of individual reporting, having successfully removed their own delusions and being intimately aware of their own mental processes.

And then there is the double standard. Scientists doing field studies are accumulating what would be called "anecdotal evidence" if it were done by members of the out-group or on out-subjects. "Scientific discovery" is done in the field, studying things in the real world such as hurricanes, volcanos, and animals in their natural environment, and discovering things such as fossils and new species. Definitions of "science" by in-group members designed to exclude the out-group and out-subjects would actually exclude much of what they consider legitimate science, if it were not for the double standard.

The other problem in having people who have succeeded at psychotherapy report their experiences is confidentiality. The whole process is supposed to be confidential. Going to a psychotherapist carries with it a social stigma. We don't want to be stigmatized by other people. I would turn that around: I would say that people who want to stigmatize other people must have some kind of psychological problem, such as the need to play God. I would ask them if THEY have been examined by a psychologist. We don't need to respect the people who stigmatize, any more than we respect those who have racial, religious, and class prejudices.

But meanwhile the social stigma exists. My psychiatrist warned me not to tell my employer that I was in therapy. One day when my boss wanted me to work overtime and there was a conflict with my psychiatric appointment, I felt I had to tell him. He said that psychotherapy was "The greatest technological development since the cave man." And he reimbursed me the cost of missing my appointment. But not all bosses are as tolerant.

I had trouble once getting a security clearance when I admitted that I had been to a psychiatrist. But then my security clearance was instantly approved when I changed my address from Provincetown to Cambridge, as if suddenly I had become less of a "weirdo."

I see all these popular books advertised, offering things such as self-esteem and peace of mind. I don't need those books. I already have all these things, as a result of psychotherapy. In the face of all the bad publicity there has been for psychotherapy, here is a list of what I have gained as a result of psychotherapy:

- happiness
- a better sex life
- greater self-esteem
- expanded consciousness
- peace of mind
- more relaxed, less rigid
- greater freedom from internal prison bars
- a greater ability to differentiate, or know the difference
- less manipulative, and less vulnerable to manipulation
- less suggestible, and less vulnerable to hypnotic suggestion
- less dependent on authority figures, and less controlled by authoritarian methods
- creativity
- intuition
- will
- compassion
- altruism
- awareness of a spiritual reality

I would think that other people who like myself have successfully completed psychotherapy would like to come forward and tell about their experiences. To preserve confidentiality they can use pseudonyms for all real-life identifications, such as name, family, location, and job. I am sure that therapists could implement this by asking for volunteers in a way consistent with ethical practices.

Every normal person can benefit from psychotherapy. When I say "normal," I am eliminating the 5% on the ends of the bell-shaped curve — those who don't have significant problems and those whose problems are so bad that they probably can't be helped by therapy. The 5% is not determined by intelligence or education or devotion to God. Everybody needs to go to a psychotherapist, just to find out if they need therapy.

Actually the way the field operates now, psychotherapists have more than enough work to do just dealing with people on the low end of the scale. But many people similar to me have been through therapy as part of their education. The New School in New York actually offered psychotherapy for credit in the 1960s. The methods and disciplines of talk therapy actually work <u>better</u> with people with less serious problems, according to a 1971 study by Lester Luborsky and associates. Other factors identified in that study as contributing to the success of therapy were high intelligence, high social achievements, and high anxiety and/or depression (or awareness of pain), which I interpret as high motivation and a high level of awareness. A 1972 study by Kenneth Howard and David Orlinsky further identified what they called "Type One Clients" as mostly young, attractive, relatively affluent, verbal, college-educated, culturally sophisticated, unmarried, and female (Lazerson, 1975, page 593). These are the kind of people who might sign up for a philosophy course. Why they are predominantly female, I don't know.

And then, in 1973, Freud was "out."

CHAPTER 12

How the Collective Psychological Defenses of an Entire Culture Rejected Freud and Jung

The first thing that anybody needs to know about psychotherapy is about psychological defenses. They are so subtle that you don't even know they are there, and this subtlety is the main source of their power. Psychological defenses are what makes psychotherapy so difficult.

The first of the psychological defenses is "I don't have any psychological problems." The same mind that has the defenses is doing the diagnosis. What's wrong with that?

You can't see your own psychology any more than you can see your own face, without a mirror or a picture. It is like saying, "I don't have a crooked nose," when everybody else can see that you do. You need someone else to keep your face looking at the mirror, while your psychological defenses make you want to look somewhere else.

It costs money to have that other person keeping your face looking at the mirror, and that is one of the most popular defenses. But actually, when you compare it with the cost of an education at Princeton, Harvard, Yale, or Stanford, it isn't that expensive. My psychotherapy, which lasted about four years, I felt was equally as valuable to me as my four years at Harvard, and cost about the same amount. Now, the same 352 hours of therapy that I had, at a top price of $200 an hour, would cost $70,400. That's just a little more than it would cost for one year at any of those prestigious institutions (and many others). A single person with a good job can afford it, as I did.

Another defense is, "Of course the therapists say you need therapy; that's how they earn their living."

But I am not a therapist. I am not making any money from psychotherapy or from promoting psychotherapy. I am only a satisfied customer. I wish that other satisfied customers would come forward and testify that a real benefit has been received for real money paid.

Another defense is to make the therapist into a bad guy. Any little flaw, or even an imagined flaw, can be used as an excuse to reject the therapist. As small children, we see our parents as perfect. The therapist is a parental figure and is expected to be perfect.

Or people avoid the real thing by seeking "alternative" therapies or reading self-help books.

Or you can fall in love with the therapist, creating a diversion.

Or you can say that psychotherapy is "brainwashing," and that I have been brainwashed.

Brainwashing is forced indoctrination using torture, drugs, or psychological stress techniques. Actually, brainwashing is more like what society has done to us, forcing us to conform to its norms. Psychotherapy is just the opposite, providing a pressure-free environment where the real self can come out of hiding after years of social pressures to conform.

The list of psychological defenses goes on and on and on. Sigmund Freud began compiling the list, his daughter Anna added to it, and people keep adding to it, because really there can be as many psychological defenses as the human mind is able to invent.

Those of us who have successfully completed psychotherapy understand the subtlety and power of psychological defenses, because we have become aware of our own psychological defenses and defeated them. So I'm sure that others who have succeeded at psychotherapy will agree with me when I say that the collective psychological defenses of an entire culture were powerful enough to get rid of Freud and the entire body of knowledge that he established.

"Psychiatrists are tools of the Establishment" was the slogan made popular by the Hippies. Actually I heard it first from an academic person in 1960, years before there were Hippies. This slogan is based on the fact that institutional psychiatrists, who are paid by the Establishment, work to serve the best interests of the Establishment. My psychiatrist, paid by me, worked to serve my best interests. As with any other occupation, psychiatrists work for whoever pays them. But once a slogan such as this becomes a hypnotic implant, people just parrot it and never examine it critically.

The Hippies were the main influence in discrediting Freud and psychotherapy. First of all, they claimed that a person could go through a complete psychoanalysis in one LSD trip. I don't doubt that this is true, in a sense, as in watching a movie. But the problem with that movie is that it is about you, and you are watching all your ego-defenses come crashing down. This was psychologically devastating to a great many people with psychological problems. You don't just jump off the ego-pedestal, as you would jump off a cliff. You climb down slowly, one hand-hold at a time.

Psychoanalysis is not a learning process like memorizing things in school. You can memorize your lines and play the hero in the movie, but to actually become that person is something else. It is a growth process, like learning to play the violin. It takes time and practice to develop every little mental circuit needed to become that physical person.

The Hippies ignored the mental in favor of the spiritual. Anything that was called "analysis" they dismissed as a "mind trip." Where I succeeded at psychotherapy by looking for faults in myself in every situation, they saw that as "negativity." They decreed that we should look only for the good in ourselves. This blocks people from the ability to learn, as I see it, because in order to learn, we must first see our errors.

LSD increased people's susceptibility to hypnotic suggestion, according to Sidney Cohen, M.D. (Cohen, 1967, page 190). This enhanced the universal unquestioning acceptance of the slogan, "Psychiatrists are tools of the Establishment." The belief that looking for faults within oneself was "negativity" further set the

emotional tone for this scathing but incoherent statement by
Alan Watts, from *The Book*:

> I am not thinking of Freud's barbarous Id or Unconscious as the actual re-
> ality behind the facade of personality. Freud, as we shall see, was under the
> influence of a nineteenth-century fashion called "reductionism," a curious
> need to put down human culture and intelligence by calling it a fluky by-
> product of blind and irrational forces. They worked very hard, then, to
> prove that grapes can grow on thornbushes. (Watts, 1966, page 11)

If this statement had been created in put-down school, I
would give it an 'A+.' It masterfully combines smear words,
truth, falsehoods, and nonsense to say really nothing comprehen-
sible. But it certainly sends a very clear social message rejecting
Freud.

Thus a whole generation was steered away from psychotherapy as education and towards more so-called "spiritual" pursuits
such as yoga and meditation.

Freud was "in" for almost three quarters of a century, and then,
suddenly, in 1973, Freud was "out." This sudden reversal might
be explained psychologically as people having been foolish
enough to believe every word he said, and then becoming seriously disillusioned when they realized that he had made some
serious mistakes. But that was no reason to reject everything he
had said and for psychologists to stop studying his works. "Freud
did bad science," they said.

Actually, the total acceptance of Freud was another example
of the childish belief that somebody has all the answers, and the
total rejection was the way that that same childish mentality
would react from the disillusion of discovering that that father-
figure, who has to be perfect in the child's eye, wasn't really perfect. The emotional shock of the disillusionment, combined with
the fact that the child's mind has not developed a very good ability to differentiate, obscures the bad logic here: Just because
Freud made serious errors doesn't mean that everything that he
said was wrong. His major discoveries, which I listed at the beginning of the previous chapter, and which I have replicated in
my own experience, still stand as major cultural innovations, as
important as the use of fire or the wheel or nuclear weapons. But

large numbers of human beings weren't ready for Freud's insights, and we have had a major cultural backlash.

If you want to see "bad science," you can look at some of the arguments trying to refute Freud.

Somebody on the Internet told me that Freud was "brilliantly refuted" in 1961, but didn't tell me by whom, or in what book. I assumed he meant *The Myth of Mental Illness*, by Thomas Szasz, which was published in 1961. A person I knew with definite psychological problems had already informed me of this book, with great delight, as his salvation from the threat of psychotherapy.

Right away the smear word "myth" signals what kind of book this is going to be, and Szasz confirms this right away by linking psychiatry to alchemy twice and astrology four times in the first two pages. "By page 6," I wrote in my notes, "I am so deep into the twisting and warping of words that I have to stop."

On page 5, Szasz says, "The psychoanalytic theory of behavior is, therefore, a species of historicism." He then defines "historicism" as "Historical ... events are viewed as fully determined by their antecedents ..." He then goes on to say, "This unsupported — and, I submit, false — theory of personal conduct has become widely accepted in our present day." I am sparing you diversions into Popper, Plato, Nietzsche, and Marx. Szasz is processing psychoanalytic theory by using the abstraction of "historicism," which he defines in such a way as to make psychoanalytic theory false.

This book could be used in undergraduate courses in critical thinking as a good example of illegitimate arguments. I plowed through most of it. It is certainly brilliant. I'll give him credit for that. But it does not refute Freud. It is convoluted and confusing. There are many instances of the kind of devious logic that appears on page 5. The word that popped into my head to describe this book is "obfuscation." It does the opposite of shedding light upon the subject — it sheds darkness upon it.

The scholars who wrote the textbooks seem to have found a clear and coherent message in what Szasz was saying. Here is one example:

In *The Myth of Mental Illness*, Thomas Szasz suggests that most mental disorders involve *problems in living*, problems that in the final analysis are solvable only by the person who has them. They cannot be cured in the way that a medical doctor cures a disease. (Lazerson, 1975, page 543)

This is what *I* believe, but did Szasz say that? Most of what I saw in this book was just an attempt to discredit the thing that I call "psychotherapy."

He goes on, in *The Myth of Psychotherapy*, to discredit psychotherapy even more:

> ... ostensibly scientific activity — rationalized ... pontificate ... pure rhetoric ... ostensibly scientific languages ... anti-individualistic, and hence threats to human freedom [page 19] ... Freud and the psychoanalysts and psychohistorians he spawned as base rhetoricians. ... Hitler's ... big lies [page 21] ... The basic ingredients of psychotherapy are religion, rhetoric, and repression [page 25] ... (Szasz, 1988, pages 19-25)

I am not really taking these statements out of context. They are the context. Those of us who have successfully completed psychotherapy know that he is misrepresenting the process in a derogatory way. It doesn't seem that he knows what he is talking about. As a psychiatrist, has he been through therapy himself? Here is his statement on that aspect of psychotherapy:

> ... All this decency and wisdom have been cast aside in modern psychiatry and psychoanalysis, which are animated by the despicable totalitarian principle that if something is bad it ought to be forbidden and if something is good it ought to be compulsory. How else can we account for the flourishing of coercive psychiatry, the famous compulsory training analyses of psychoanalysts, and the massive medical, psychiatric, psychological, and legal apologetics written on behalf of such compulsory soul-curing? (Szasz, 1988, page 37)

I wouldn't go to a therapist if I didn't think that that person had been through psychotherapy and succeeded at it, any more than I would accept as my guide to climb Mount Everest somebody who had not successfully climbed it. As for the despicable compulsory training, how did he get his M.D. degree except through compulsory training? I am wondering what motivated this strong propaganda piece? What is his personal experience with psychotherapy? He doesn't say. He talks about Luther.

Another propagandist against psychotherapy is Frederick Crews. He raises the question of whether "only the analyzed can judge" (Crews et al., 1995, page 8). Like Szasz, he gives no indication that he has experienced psychotherapy. In fact, he comes right out and says, "... My qualifications for making pronouncements about psychoanalysis and psychotherapy without benefit of analytic training or even of a personal analysis ..." (Crews et al., 1995, page 6). He seems to have studied works on psychotherapy by philosophers and other intellectuals.

Crews is very, very intelligent. I am not intelligent enough to argue with Crews. He knows a great deal, yes, but, like Szasz, he doesn't seem to know very much about psychotherapy and the way it works. So is he really saying anything about psychotherapy?

E.M. Thornton, in *The Freudian Fallacy: Freud and Cocaine*, is trying to invalidate psychoanalysis by pointing out that Freud was using cocaine at the time he formulated the idea. On the cover is an artist's creepy distortion of Freud's face, and inside is more of the same – a portrait of Freud in a very unfavorable light. Even if it is all true, I would call it an "ad hominem" argument, because psychoanalysis is not invalidated by making Freud look bad.

These people and others are taking advantage of the huge demand for literature supporting people's defenses against psychotherapy:

> ... And there is such a ready audience for publications that speak ill of psychotherapy that potential debunkers can be published when they have no data at all to substantiate their claims. ... (Smith et al., 1980, page 47)

Carl Jung was rejected by the academic establishment long before Freud, maybe 50 years earlier. I once mentioned the name of Carl Jung to a scientist I knew, and he said, "Carl Jung was a mystic." I couldn't imagine what he was talking about. A mystic is a person who attempts to bypass the intellect and the evidence of the physical senses — just the opposite of science — in trying to perceive or experience a deeper reality. To call Jung a "mystic" is a gross misrepresentation. I thought my friend was using the second definition of "mystic," the derogatory definition, meaning "illogical" or "irrational." But since then I have seen the word "mystical" applied to Carl Jung in a textbook:

... One of the most mystical and metaphysical of psychological theorists, Jung has had greater acceptance in Europe than in America. The popularity of his works, however, appears to be growing in America, especially among persons interested in mysticism and nonempirical aspects of psychology. (Lazerson, 1975, page 423)

I know that Carl Jung was mostly rejected by the academic community in the beginning, and that he has gradually gained acceptance over the years. But why call him "mystical?" They don't explain what they mean by this. And why do they say "nonempirical?" Jung's discoveries were very much based on evidence, although, as with Freud, much of this evidence came from internal perceptions, and mostly from dreams. Carl Jung, actually, was more of a scientist than those who rejected him, because when he saw evidence of a spiritual reality in his patients' dreams, he accepted it.

But, as I see it, an academic in-group that had to conform to a purely physical view of the universe had to find a way to say "not one of us," and the way they chose was to identify him with the out-group, by calling him a "mystic." This is a common ingredient of in-group jokes and insults. And so there is this widespread acceptance in the academic community of this labeling of Carl Jung as a "mystic," without a peep of protest, because everybody in the in-group understands the joke.

So Carl Jung is here bundled in with people who reject reason and the evidence of the senses, and spend most of their time meditating. And he is set up to be dismissed as "mystical," and therefore "unscientific." How long will the culture live with this kind of inaccuracy?

I just happened to come across the following, from an interview with Jungian analyst Marie-Louise von Franz about her former relationship with physicist Wolfgang Pauli:

[von Franz]: Pauli was afraid of the content of his dreams. It frightened him to draw conclusions from what his dreams said. They said, for instance, that he should stand up for Jungian psychology in public. And that he feared like hell. Which I understand. He moved in the higher circles in physics. They were very mocking and cynical and also jealous of him. If he had stood up for dreams and irrational things, there would have been a hellish laughter. ... (van Erkelens, 2002, pages 146-7)

We all know there would have been a hellish laughter. But "hellish laughter" doesn't have anything to do with science. "Hellish laughter" is part of the bullying that children learn to do in the schoolyard.

The Hippies bypassed the mental in favor of the spiritual. Psychiatrists were tools of the Establishment. Finding one's repressed evil was negativity. The Hippies were above all that. They were spiritual. They were superior.

The Hippie influence led younger generations into spiritual pursuits such as meditation and yoga. M. Scott Peck was very successful with *The Road Less Traveled,* calling psychotherapy "spiritual," but for the most part psychotherapy was "out." Psychotherapy was for those less fortunate, to bring them up to normal, but not for spiritual advancement, as it had been for the "Type One Clients," before LSD.

Although I can't argue that psychotherapy should replace spiritual pursuits such as yoga and meditation, I have always felt that it should play some complementary role. Spirituality works best in a mind that is free of psychological problems. And I have seen signs that not every "spiritual" person is free of psychological problems.

I have always seen the attitude of superiority as ego-compensation for feelings of inadequacy. I have known several people involved in New Age pursuits such as meditation, yoga, and spirit guides who had obvious psychological problems and sought psychological help. I have also known two women who told me that their yoga instructors had tried to rape them.

Then one day I was having a great conversation in the office with a woman whose partner happened to be a yoga instructor. Then he showed up, apparently having found out about our social intercourse by his clairvoyant powers, and paralyzed my mind, so that I wasn't even able to resume the conversation after he left. It was a remarkable show of "powers," but also a remarkable show of childish possessiveness. She was already planning to leave him and move across the country.

So yes, I concede to the "powers," but the powers weren't getting this man what he wanted — his woman. Some psychotherapy might have helped. I am not saying that psychotherapy defeats the "powers," but only that psychotherapy is a complementary discipline to get you where you want to go. And without it, you won't get far on your spiritual journey. You will be wasting your powers on childish pursuits.

I was trying to find ways of communicating this idea (past all the defenses) when I discovered "spiritual bypassing" on the Internet. I was looking up "psychological defenses" when I found the article "Beware of Spiritual Bypass," by Ingrid Mathieu, Ph.D., identifying spiritual bypassing as a defense mechanism. What it means is that people who are involved in spiritual pursuits when they still have unresolved psychological problems should be working on resolving their psychological problems instead. Of course, being "spiritual" is seen as being "superior," and having psychological problems is seen as being "inferior," so spiritual bypassing works perfectly emotionally as a psychological defense, giving people the impression that they are involved in the superior activity, and that surely it will give them things like self-esteem and peace of mind, when actually these things would be achieved better by psychological methods.

The term "spiritual bypassing" was originated in 1984 by John Welwood, a Buddhist. He explains how nonattachment is really impossible if one is attached psychologically to unresolved traumatic experiences of the past. In an interview with Tina Fossella, he says:

> TF: *You introduced the term "spiritual bypassing" 30 years ago now. For those who are unfamiliar with the concept, could you define and explain what it is?*

> JW: Spiritual bypassing is a term I coined to describe a process I saw happening in the Buddhist community I was in, and also in myself. Although most of us were sincerely trying to work on ourselves, I noticed a widespread tendency to use spiritual ideas and practices to sidestep or avoid facing unresolved emotional issues, psychological wounds, and unfinished developmental tasks.

> When we are spiritually bypassing, we often use the goal of awakening or liberation to rationalize what I call *premature transcendence:* trying to rise above the raw and messy side of our humanness before we have fully faced and made peace with it. And then we tend to use absolute truth to disparage or dismiss relative human needs, feelings, psychological problems, relational difficulties, and developmental deficits. I see this as an "occupational hazard" of the spiritual path, in that spirituality does involve a vision of going beyond our current karmic situation. (Welwood & Fossella, 2011)

I very much recommend reading the whole interview, because it explains important differences between the psychological part of spiritual growth and what lies beyond.

The process that works its way into all these defenses is rationalization — inventing a seemingly rational argument to support whatever your emotions want you to believe. I define "rationalization" as an argument that is not quite accurate and not quite valid, but is accurate enough and valid enough to convince somebody who really wants to believe it.

All these various defenses, at all levels, have kept individuals and the entire culture from pursuing psychological wellness as part of their educational and developmental experience. Yes, Freud made mistakes, but Freud was the pioneer. And perhaps the cocaine experience was necessary for him to blast through the total hypocrisy of 19th-century Western civilization and achieve his view of self-knowledge and how to attain it. Self-knowledge was a radical enough concept in itself, in a time when people were trained only to look outward at an external reality. Yes, Socrates said, "Know thyself," but that really had no meaning in a time when people didn't really know that they had a "self." And yes, psychological problems did all have to do with sex, or mostly, anyway, in a Victorian Era when sexual activity was severely repressed and regulated. So, despite his many mistakes, Freud made a major breakthrough into a whole new cultural understanding — ranking, as my boss said, with the discovery of how to use fire or the invention of the wheel.

Carl Jung carried Freud's work farther, correcting some of his mistakes, and recognizing that there was a spiritual dimension to human existence, contrary to in-group beliefs. Labeling him a "mystic" is a major cultural error. Dismissing him as a "mystic" is an atrocity.

We, the "Type One Clients" of the early 1960s (before the Drug Revolution) acquired a huge body of knowledge from our psychotherapy, about psychological problems and psychological defenses and ulterior motives. By abandoning the discoveries of

Freud and Jung, the culture is losing an important body of knowledge.

CHAPTER 13

Psychotherapy as the Way to a New Civilization

Why is Freudian/Jungian depth psychology important? Isn't it out of date?

It has gone out of fashion with people who prefer more "spiritual" pursuits for their personal growth. But as I have already pointed out with "spiritual bypassing," one's childhood traumatic experiences must be resolved before one is free to concentrate on more spiritual disciplines. More than that, psychotherapy can be a spiritual journey in itself, as was explained by M. Scott Peck in his enormously successful book, *The Road Less Traveled.*

But he wasn't aware, and most psychologists aren't aware, and the culture in general isn't aware of the discoveries I made that showed me how psychotherapy can lead us to a new civilization. After I had successfully completed psychotherapy to the satisfaction of my psychiatrist, I spent ten more years actively working on my psychological development, using dream analysis and free-expression "scribbling," until I, myself, was satisfied that I had reached psychological adulthood. In that process I discovered how "human nature" itself could be changed, to create a society that could truly be called a "civilization," along with a few other things that aren't generally known. My experience is described in my book, *Re-Educating Myself.* But because my discoveries are not part of the cultural awareness, I feel that I must take a few pages to describe them, to show where our culture might go if it was not blocked by the invalid arguments of physicalism and psychological defenses. To start with, I think I need to explain what I

101

mean by a "new civilization" and why we need such a thing.

In the summer of 1953, when I was 19 and had just finished my freshman year at Harvard, the Soviet Union exploded their first hydrogen bomb. Suddenly the two major powers had both the motive and the means to destroy each other and threaten all of human life on earth. Within five years, the word "overkill" had to be invented, to mean that both sides had many times the destructive power necessary to destroy each other and probably exterminate the human species as well.

Suddenly warfare became unthinkable. The manly virtues of fighting and warfare, which had been instilled into every boy child since before recorded history, and had been instilled also in every girl child insofar as she learned to respect and admire the men who protected her, suddenly ceased to become virtues and became negative forces that could destroy us. We needed a whole new way of thinking.

In 1954, when I was 20, I was thinking about what I wanted to do with my life. My father and grandfather before me had worked to make money. I already had the benefit of that money and social position, but I wasn't happy. This may sound like a personal problem, but it isn't. I was the end product of the American dream. The promise of America is that people are free to work their way up to a level where they have money and material comfort, if not for themselves, then for their children or grandchildren. The implied premise in this is that it will bring happiness. Certainly it will bring happiness compared to living in poverty. But indoor plumbing, electricity, central heating, radios, television, cars, boats, summer places, money in my pocket, and social status did not make me a happy person. I had to abandon the American dream, not because it had failed, but because it had succeeded. There was more to happiness than it offered. I had to find a new "standard of living." Also, because I had already reached the goal of the American dream, I had to find something else to do with my life — new goals and a new way of life.

It seemed to me that the happiest people were those who had developed their full potential as human beings, regardless of what that potential was or whether they were "successful" by cultural standards of money, power, and fame. So I set out to develop my human potential, in search of happiness.

The following year I studied "The Waste Land" by T.S. Eliot. This had been presented to me in two different English courses as an important piece of twentieth-century literature, but also as apparent madness, as in "This won't be on the exam." I wanted to resolve the question of why, if this was apparent madness, was this an important piece of literature.

It was a depressing experience. The old religious beliefs from which we had derived our values had collapsed, leaving twentieth-century humanity stumbling amid broken images, living boring and meaningless lives and working at boring and meaningless jobs. After about three weeks, I finally understood why it was so depressing: "The Waste Land" was my world. I was living it.

I looked to T.S. Eliot for a solution to the problem, but all he offered was: "Shall I at least set my lands in order?"

That wasn't a solution. I was the model boy, much admired and respected. I was fine. It was the world that was a mess.

But that was the only solution he offered, and the more I thought about it, the more I could see that it was a solution, and perhaps the only solution. Efforts to change other people would result only in resistance from them, unless they wanted to change. But if I wanted to change, I could change myself, and change my small fraction of the world. This was also consistent with my goal of developing my potential as a human being.

In 1955, at age 21, because of these cultural problems, especially the threat of nuclear annihilation, which I saw as total systems failure of the culture itself, I turned my back on the culture, as represented by the seven million books in Widener Library, and set out to design a new civilization. "New civilization" was the best expression I could think of to mean new values, new beliefs, new ideals, new goals, a new "standard of living," a new way of thinking, and a new way of life. A civilization is not just buildings and cities and large numbers of people, but it is based on a set of fundamental ideas that determine what those people choose to build and how they choose to live their lives. The "civilization," to me, was what had been programmed into my head as a product of the American upper middle class. To create a new civilization, all I had to do was change that programming. Of course this is much more easily said than done.

In 1956, my senior year at Harvard, I applied for Navy Officer Candidate School (OCS). This had been my plan all along, to avoid being drafted in the Army. But also, semi-consciously, I was using their comprehensive and stringent testing as a kind of self-evaluation.

I passed the dreaded physical exam, which many a star athlete had failed because of some minor physical disability. They told me I was qualified to be a fighter pilot.

On their Mental Exam, they told me I had received the highest score they had ever seen at their Boston Headquarters. This was in competition with other students from Harvard, MIT, and all the colleges in the Boston area. This answered my question: Who am I to be designing a new civilization? Nobody I knew of seemed to be qualified by virtue of education or experience, so that made it a question of raw mental ability. And my raw mental ability had just been acknowledged by the culture.

In all fairness, the Navy test was biased in my favor. In addition to the usual "mathematical" and "verbal" sections, it had also a "mechanical" section, which tested one's knowledge of boats and the maintenance of boats. I know I got a perfect score on that section, because I had already worked summers on a sport-fishing boat, in a boatyard, and as a professional sailing master. But the test told me what I wanted to hear at the time.

But on the Medical History Questionnaire, I answered "Yes" to the question, "Are you often depressed?" This led to an examination by a psychiatrist, who diagnosed me with "mild chronic anxiety" but said that this would not prevent me from performing my duties as a Naval officer. (A friend quipped that a little anxiety was just what the Navy wanted in a junior officer.) But because the class was oversubscribed, I was rejected because of this "abnormality."

Wiser heads told me that I should not have answered "Yes" to that question. They didn't know that I was using the Navy's screening process as a self-evaluation.

I had been turned down for a job because of a psychological condition. The intellectual part of me had been educated, but the psychological part had not. So I saw psychotherapy as a necessary part of my education. And as it turned out, psychotherapy also led me to a new civilization. In hindsight it is obvious that it would, in a culture where most people have an intellectual education but not a psychological education.

Actually my psychotherapy started with my study of "The Waste Land." This was the hardest thing I had ever done in my life, at the time, because to understand "The Waste Land," I had to be able to see myself and my life. We are trained in our intellectual educations to look outside at external objects, but not inside at ourselves. It is as hard as seeing one's own face, without a mirror. I remember being taught in school that Socrates said "Know thyself," but what that meant was never explained. It was just something we had to memorize. Self-knowledge was simply not part of the American culture of 1955.

I learned to look upon myself from "on high." This is not out-of-body travel, but simply separating my mind from my ego and seeing myself as a person "down there" doing foolish things and struggling with the problems of life. This was a good warm-up exercise for psychotherapy. Even so, psychotherapy was the hardest thing I ever did in my life, at the time.

I have said, in two published books, "Psychotherapy is the way to a new civilization." That isn't exactly true. Drug therapy isn't the way to a new civilization. Also I think that most psychotherapy only brings people from subnormal to normal. A new civilization is something beyond normal. Also, not everybody succeeds at psychotherapy. In order to succeed at psychotherapy, and to reach a level that I call "a new civilization," I had insights that I didn't read about in books. So it would be more accurate to say, "Psychotherapy AS I EXPERIENCED IT is the way to a new civilization." I want to share my experience with psychotherapy, with emphasis on those aspects of it that were different from the prevailing cultural view.

First of all, I want to define "psychotherapy," as I experienced it and as I use the word in this book: Psychotherapy as I experienced it means resolving one's childhood traumatic experiences and growing to an adult level of behavior. Historically this has been achieved by talk therapy, as differentiated from drug therapy, or psychopharmacotherapy, but there is always the possibility of new and more efficient methods being invented. Of course there is a whole range of psychological problems that professionals have to deal with, starting with outright insanity, but I am focusing here only on the relatively mild disorder that I experienced, insofar as it is representative of the condition of the normal person.

I define psychological "wellness" in terms of "appropriate behavior." "Appropriate behavior" means behaving in such a way as to achieve the desired results. It means channeling the raw energy of one's physical drives through one's mental circuitry in ways that improve as one grows older and wiser. You might think of it as programming, as creating software for living. It is the supreme creative activity, the creation of a human being, living up to one's adult potential.

A perfect example of appropriate behavior is that of Martin Luther King, Jr. When provoked by police brutality, he didn't try to fight back, and he didn't submit to physical domination, either. He did exactly what was needed to succeed at winning his civil rights.

People have been debating whether psychotherapy is science, or art, or religion. I see it as education. I had had 19 years of intellectual education, but psychotherapy was emotional education, to develop the emotional circuitry in my brain/mind to produce appropriate behavior. It is more closely related to philosophy than to medicine.

The child, with the child's mind, is confronted with an overpowering situation with powerful emotional content — fear, pain, and/or anger. The child doesn't understand — doesn't know what to make of it. The child has an unresolved problem in living — in philosophy, if you will. The problem is filed away in the child's mind (consciously or unconsciously), and into adulthood the person is driven (usually unconsciously) to keep recreating the same situation over and over again until the problem is resolved. This is the "traumatic experience" and the resulting "fixation," as Freud defined it, and as I know it.

As part of recreating the same situation over and over again, the person must preserve the same childhood self — at least the same attitudes and level of psychological development. This is the "inner child" that everybody knows about.

So the whole idea of therapy is that the adult with the adult's mind has a better chance (not a sure thing) of solving the problem than the child with the child's mind, if only he/she could remember the details.

Mental disorders were described in the mid-20th century as "mental illness," to get away from the stigma of demonic possession: It wasn't as if you had consorted with the Devil. It was like being sick. It wasn't your fault. But that didn't work. The stigma then became attached to being "sick."

And the idea of "mental illness" was somehow associated with the "medical model." The brilliant doctor was going to analyze you and somehow make your illness go away.

Before I began psychotherapy, I had read a great many case studies, showing how the brilliant analyst made a brilliant diagnosis. But if that was the diagnosis, then what was the cure? They never explained that.

In therapy I began to see it as a problem and a solution, rather than as a diagnosis and a cure. Once the solution is found, the problem goes away.

In the "medical model," the brilliant doctor performs the operation, and the patient is only a piece of meat to be operated on. Psychotherapy doesn't work that way.

The client must solve the problem. The client must do the analysis. The client has all the necessary information in his/her own head, although much of it is in the subconscious and jealously guarded. The proper role of the therapist is only as a guide, to recognize the tricks and traps and ego defenses, and employ whatever strategies are necessary to enable the client to solve the problem. If the brilliant analyst sees the solution and presents it to the client before the client is ready to accept it, the client's ego defenses will reject it.

I see no mention in the literature of psychotherapy that the client has to solve the problem. I think that this one misunderstanding is the main reason that psychotherapy usually fails. First of all, the client expects the doctor to solve the problem. Second, the client treats the doctor as an authority figure, as in "I can fool the shrink any time I want to." And third, the client senses on some level that he/she is being exploited to support the therapist's ego props, when the therapist takes the credit for success. This creates resentment and therefore a bad working relationship.

This can actually be a traumatic experience in itself.

There are vestiges of authoritarianism here. The therapist is not the authority, like a teacher or a doctor. The therapist is only a guide, keying on the client's defenses, so that the client can find his/her own way to his/her own inner truths. The therapist is not the idealized parental authority figure, although the client, who is still psychologically a child, will put the therapist in that role, and the therapist, if not psychologically mature, may be playing that role without actually being aware of it.

I think psychotherapy would be successful for many more people if the clients understood, and if it was part of the cultural awareness, that it wasn't the therapist's job to solve the problem, but that it was the clients' job to reach deeply within themselves and find their true selves.

When I first began therapy, I was sitting one day trying to find my real self, and the self I found was my four-year-old self. This was the self that I identified with, that I felt was the real "me." So of course I was uncomfortable and anxious, trying to relate to adult women and trying to find a job in New York City.

But being four years old psychologically did not make me freakish in the eyes of the outside world. I appeared normal. I had graduated from Harvard, had served honorably in the Army, and was hired out of 75 applicants for my first job, at which I was successful right from the beginning (still psychologically at age 4).

But deep inside of me, there was this child I had been hiding for years. And once I recognized this child, it was a great relief. He was free — free to exist within my own mind, and free to develop and grow up to become an adult.

I have asked a few people if they could find a childhood "self" that they identified with, or a time in their lives when they really felt they were a real person. Two of them identified themselves as age 11 and 8 respectively. This is within the normal range, as I shall explain later. Psychological age 4 is below the normal range, but this was still not conspicuous in my case, as I have said. So I am wondering whether you can find a "real self" that you can identify with at some childhood age?

The idea of the "inner child" was "in," and then became a cliché, and then was ridiculed by people who were embarrassed to admit they had an inner child. Don't let those other children intimidate you. The concept of the inner child is important, and serious, and must be preserved.

Let's move this up to a higher intellectual level. I remember people using the word "existentialism" during my college years and beyond, roughly 1954 to 1964. They used it to show how intellectual they were, but when asked to define it, no two people ever gave the same definition, and I don't think any of them knew what the word meant. This is what I think it means: I think it was 20th-century humanity struggling to make the assertion, "I exist." It was the first assertion that such a thing as a real self even existed. The Jane Austen novels are wonderful documentation of how totally artificial people's lives were in an earlier century, totally dictated by social roles.

The child is told to share his/her toys with other children, and told that this is "good." The child doesn't want to share his/her toys. So the child believes he/she is inadequate and abandons his/her real self for an artificial self, and beyond that point begins playing a role that society has scripted for him/her, like playing a part in a movie, being everything the Establishment wants him/her to be. Just the opposite of "Psychiatrists are tools of the Establishment," psychiatrists can free you from being a tool of the Establishment.

Existentialism was an attempt to break out of this total artificiality and assert that a real person existed in there somewhere. Existentialism can't really be explained intellectually. It has to be experienced, by finding that real person.

"Intellectualism" has been defined as a psychological defense. I don't know that it is so much a defense as a way we have been trained to use our minds, after many years of intellectual education. Discussing the real self intellectually won't find it. The real self is found by looking, and feeling, to find out if it feels real. My psychiatrist kept asking me, "What are your feelings?" Most of the time I had no feelings, actually. But feelings are the key to knowing whether it is a real self or an artificial self that one is looking at.

With intellectualism comes abstract language, such as "existentialism," actually. Abstract language is always ambiguous, because everybody has their own unique set of specifics that go into creating the abstraction. Many people's definitions of "existentialism" will differ from mine.

To avoid the problems of abstract language, my psychiatrist insisted that I speak in specific language, describing specific actions and thoughts — not "I had an existential experience," but "I found my real self at age four."

After three and a half years, my psychiatrist told me I could end the treatment. He thought I was OK. I was normal. I felt normal, although I felt that there was still something missing. I stayed on for another year, because I felt dependent on him, until my company moved me to California.

A year later I was back East, in Provincetown, which had been my family's summer home for generations, ostensibly to write my philosophy, but actually, in the solitude of Cape Cod winters, I was able to build a philosophy. I experienced "withdrawal," as defined by Arnold Toynbee in *A Study of History* and experienced by Henry David Thoreau on Walden Pond and Isaac Newton at the family farm, where he observed the falling apple and formulated the law of gravity. In the solitude of Cape Cod winters, away from the cultural din, I was literally able to hear myself think, and this became the greatest creative period of my life.

I had to learn about Carl Jung from a Provincetown artist, because Jung was hardly mentioned in all the psychology books I had read. In December 1966, with the help of my nine-pound dictionary, I read *The Basic Writings of C.G. Jung*. I asked myself, "Where did he find his archetypes?" And the answer was, "in dreams." So I set out to explore my dreams in search of archetypes.

But the dreams had an agenda of their own. They picked up where my psychiatrist had left off. They showed me that I was still only 10 years old psychologically, and that this was normal for the culture.

The discovery that the normal person is psychologically 10 years old can be, and should be, as important to the culture as Newton's discovery of the law of gravity. But I'm sure that there are academic people with PhDs who are psychologically 10 years old, who will fight it with all the ingenuity that their psychological defenses can produce, at least when they can no longer ignore it.

I thought that when I became psychologically normal, I was psychologically an adult. But the dreams were telling me that this was not so, and just to reinforce the message, there were some 10-year-old boys in real life who wanted to play with me.

So here is another confusion in the field of psychology: The word "normal" means average, in the usual sense, and the word "normal" also means psychologically well. But these are not the same thing. Yes, "normal" means that a person is not insane, but "normal" is a long way from being psychologically mature or reaching one's true adult mental potential.

The good news is that the average adult in the United States is operating with only his/her 10-year-old mental potential and has something to look forward to in reaching his/her true adult mental potential and the quality of life that goes with it.

But the psychological profession does not currently recognize that true adult mental potential. Their goal right now is only the halfway point, the "normal," the average. I want them to tell me that I am wrong about this.

I reasoned that if I was only 10 years old psychologically, then maybe that explained why my sex life wasn't as good as I imagined it could be. I theorized that if I could reach the psychological age of puberty, or psychological age 14, my relationships with women would improve. So, using the disciplines I had learned in psychotherapy, I set out to analyze my dreams, picking up where my psychiatrist had left off, with that goal in mind.

People don't understand what I mean by "the disciplines I learned in psychotherapy," because they think that the analyst does all the analyzing and the "subject" is only an object. But if the "subject" has to solve the problem, I am sure it can be understood that the "subject" needs to learn certain disciplines — recognizing one's defense mechanisms, recognizing one's rationalizations, being able to see oneself and recognize one's faults, learning to look for details and not just philosophizing, knowing the kinds of things to look for, and so on.

In 1966, interpreting one's own dreams was unheard of. I asked myself, "What will happen if my interpretations of my own dreams are wrong? Will I sail off into some wacko-land?" The dreams themselves answered that question.

I was having some dreams that I interpreted as homosexual, one about the dentist, the guy who puts his tool in your mouth, and another about a fire in the corporate chimney. I thought that I might be homosexual. I was adjusting mentally to an identity as a homosexual when I had the following dream:

A beautiful woman is lying on a couch naked, absolutely drooling with desire for me. As I start towards her, suddenly I am in a car full of boys — teenagers or homosexuals — driving around town and having a wonderful time. But all I want to do is get out of that car and back to that woman. Finally, with a supreme lunge, I get out of the car — and wake up.

It was such a powerful dream that for a week afterward I was trying to get back into that dream and back to that woman. That answered forever the question of whether I might be a homosexual.

Also it answered forever the question of what would happen if my interpretation of my own dream was wrong: If your interpretation of a dream is wrong, the dreams themselves will correct you. This is a logical extension of Carl Jung's description of dreams as a compensation for, or correction to, the conscious attitude. If your interpretation becomes part of your conscious attitude, then your dreams can correct it.

I am leaping here from the particular to the general because in 1991, when I presented my paper on "The Self-Steering Process" at the annual conference of the Association for the Study of Dreams (ASD), I learned that many of the dreamworkers there were familiar with the self-steering process already. What nobody seemed to know at the time, and I didn't learn myself until 2004, was that Carl Jung had mentioned it in a paper in 1931. Why it hasn't been publicized in popular dream books, I don't know, because it makes self-interpretation of dreams an accurate source of knowledge which rivals science itself.

It used to be that only the analyst was considered qualified to interpret a person's dreams. But actually, given the self-steering process, dream analysis works best if the person having the dream also interprets it, given some grasp of the necessary disciplines. I saw this reflected in the "ethics" statement for the 2013 conference of the IASD: "Workshops conducted in a manner that implies that the leader is the authority on the meaning of the dream, rather than the dreamer, are unacceptable."

Getting back to the winter of 1966-67, I had to tell the first PhD psychologist I saw about my discovery of the self-steering process, and her reply was, "I don't think They would agree with you." This is the caste system speaking. This is a person playing the role of the idealized parental authority. Actually "They" don't even study dream analysis any more, so what do They know?

You can read about my dream winter in detail in *Re-Educating Myself*. I did reach psychological age 14, the psychological age of puberty, and my sex life did improve as a result. That transition point was marked by the following key dream:

It was my first day in the Coast Guard. The waters had risen, burying man, woman, and child under 120 feet of water. It was my job to go out in a very small dinghy and mark the spot, so that other people could rescue them. There were huge 14-foot sharks in the water, and it was almost certain death to go out in such a tiny boat. There were larger boats, but I hadn't earned the right to use them. I was to be married that afternoon to the beautiful young woman I had met in the supermarket the day before. I didn't want to die, when suddenly I had so much to live for. Two older Coast Guard men stood patiently holding the boat for me, waiting for me to make my decision. As I stepped into the boat, I woke up with a jolt.

I was ready to sacrifice my life for man, woman, and child. The number 14 was a hint that this was psychological age 14.

This is the psychological age of puberty. This is a transition point in human psychological growth where "human nature" itself changes, from the exclusive self-interest of the child to an equally natural drive to give and share and sacrifice oneself for others.

This can be explained in terms of survival of the species: The survival of the species is not only "the survival of the fittest," meaning the survival of the individual, but it also involves the survival of the greatest number of individuals. Survival of the species is best served if the children, who are the most vulnerable, look out for their own interests. But when they become adults, with children of their own, it is more appropriate that they should be motivated to give and share and occasionally sacrifice themselves for the survival of their children and the greatest number of human beings.

If my discovery that the normal person is psychologically 10 years old is as important to the culture as Newton's discovery of the law of gravity, then my discovery of how "human nature" itself can be changed should be as important as discovering how to undo the force of gravity. Can the culture be changed this way, or was this true only for me? I had only the one dream telling me that the normal person was psychologically 10 years old. That is not much evidence to support extending my discovery to the whole human species. But remember the self-steering process: I have held this view as part of my conscious attitude since 1967, and my dreams have never corrected it. And you can help me by looking for the childhood "self" that you identify with.

Also my view is consistent with economic and political reality. The American system of competitive self-interest works because it is in tune with the level of psychological development of the average person. More altruistic systems such as socialism don't work, because the average person is not ready psychologically to function at that level, and the forces of self-interest dominate.

Unlike physical growth, psychological growth is not automatic. Most people's psychological growth is arrested before they reach physical puberty, and they never reach the psychological age of puberty in their lifetime. But if a majority of people were to reach this transition point from competitive self-interest to altruism, we would have a new civilization. That's how psychotherapy as I experienced it is the way to a new civilization.

We all know about altruism. Altruism is nothing new. But there is a difference between playing the role of altruism, as an actor in a movie, and actually being an altruistic person. Reaching the psychological age of puberty means actually becoming that person, to the depths of one's being.

I think our cultural ideas of morality and altruism came originally from people who reached this level of psychological development and beyond. But this level of being is not achieved by forcing children into artificial roles, but in letting them be children and letting them grow naturally to psychological adulthood.

I continued to work actively at my psychological development until I reached psychological age 18, or the psychological age of adulthood. At that point I felt totally competent as an adult and comfortable with myself, and I was not motivated to develop myself further psychologically.

Richard Kieninger has written that it is possible to keep growing psychologically to psychological age 28, acquiring psychic abilities such as controlled clairvoyance.

My concept of "psychological age" is my original creation. Richard Kieninger discovered it independently. It is not the same as Freud's or Erikson's stages, or Maslow's hierarchy of needs, or Kohlberg's stages of moral reasoning, nor is it derived from any of those systems. The idea of psychological "age" has the advantage over all those other systems in being numerical, which means that it is measurable.

Unfortunately, I don't have either the qualifications or the credentials to design a test of psychological age, so I offer the idea to anybody in the field of psychology who might be inspired to do it. The field of psychology has documented human physical development in exhaustive detail, but I see no indication that what I call "psychological" (mental, behavioral, philosophical) growth has been documented in the same way. One way of measuring psychological age would be to measure the ability to differentiate, which I found improved as I developed psychologically.

When I tried to tell a PhD psychologist about the psychological age of puberty, she said, "Maslow must have said that." Foolishly, I made it my job to find out where Maslow had said that, and discovered that there were 17 publications by Maslow listed in the Dartmouth College Library catalog. That made the task impractical. I should have put the burden of proof on her. Since she was the expert in psychology and she was the one making the assertion about Maslow, I should have insisted that she back up her statement with a book and page reference and an exact quote. But for the moment her bluff worked and she was able to brush me off.

Actually, I was being doubly duped here. Her implied assertion was that if Maslow had said the same thing that I was saying, that would give her an excuse to dismiss what I was saying, whereas actually this would support what I was saying.

Did Maslow say what I am saying? I leave it up to scholars to really answer that question. But judging from the brief description of his thinking and his approach that I have seen on the Internet, he did not, nor would he have been likely to have made the same discovery that I did.

First of all, there is no mention of a transition point where "human nature" itself changes from competitive self-interest to compassion and altruism. He wouldn't have observed that anyway, because I see no mention that he observed people in the process of psychological growth. He studied famous and distinguished people in their fully developed state. His thinking was theoretical and abstract, whereas I made empirical observations of the changes in motivation of one human being (myself), using my mental senses.

I saw the word "altruism" once, and compassion implied, in a long list of attributes of "self-transcendence." They are only mentioned. There is no mention of how important these attributes might be, towards his vision of world leaders sitting at a peace table. Instead, he bypasses these attributes and emphasizes "peak experiences" as the most important attribute of the highly developed person. I have also seen "cosmic consciousness" mentioned. These things bring to mind the psychedelic drug experience, not the gradual growth process experienced in psychotherapy. The whole idea of "self-transcendence" implies going beyond "self," which again is reminiscent of the loss of ego experienced in an LSD trip. Psychotherapy is a process of strengthening the ego, or the real self, as differentiated from the ego compensation, or the psychological defenses of the artificial self. The real self is never transcended. So I would say that Maslow was much more aligned with the Hippies and their psychedelic drug experience than with anything I am saying about psychological growth and maturity. They would say that their view is superior, but I would say that my view is more likely to lead us to world peace and survival of the species.

Also the whole idea of "needs" implies something consciously felt and some conscious motivation to do something about it, whereas psychological growth, such as the transition experienced at the psychological age of puberty, happens without any conscious idea of what it is going to be. For example, I had no idea that my selfish desire for sexual pleasure would be satisfied by attaining attributes of compassion and altruism.

So the representation of Maslow on the Internet gives no indication that he might have discovered anything like the psychological age of puberty, or even that his thinking and approach would have led him to make such a discovery.

And with this long discussion of Maslow, I see that I have been duped in yet another way: Getting involved with Maslow has created a diversion which distracts attention away from what I am saying. Whether or not Maslow might have said the same thing makes no difference to the accuracy and potential cultural value of what I am saying. So let's get back to what I am saying:

- My dreams showed me that the average person is using only his/her 10-year-old mental potential.

- Continuing my psychological growth beyond normal, I experienced a transition point where my "human nature" actually changed, from a basic motivation of competitive self-interest to an equally natural motivation of compassion and altruism.

- This is not "theory" but is empirical evidence, observed by me through my mental senses.

- I am a sample size of one, not significant in itself, but waiting to be combined into a significant sample through meta-analysis.

I am hoping that persons with scientific credentials will recognize how important this discovery would be if it were true, instead of just trying to dismiss it.

Getting back to "overkill," nothing in our Western culture prevented us from entering into an arms race and following it through until we had developed the minimum destructive power necessary for the extinction of our own species. I say "minimum" because certainly our technology will develop more powerful, more efficient, and more certain ways of making our human species extinct. And I learned, after I had been working on the problem for years, that I was doing something that used to be called "philosophy," before Western philosophy deteriorated into some kind of useless intellectualism. I say "useless" because Western philosophy has not, to my knowledge, provided us with any solutions to the problem of imminent extinction.

Well, actually, British philosopher Bertrand Russell sent off telegrams to Kennedy and Khrushchev and other world leaders during the Cuban missile crisis in 1962. He might have prevented World War III, but with diplomacy and common-sense appeals to reason and restraint, and not with anything out of Western philosophy.

But Sigmund Freud developed a method that could ultimately save us, a few years before Einstein published his formula that could ultimately destroy us. Of course the culture has no idea how that is possible. It is not known that normal people are functioning on only their 10-year-old mental potential, that their "human nature" itself would change to compassion and altruism if they were to function with their 14-year-old mental potential, and that if majorities in the major powers were to reach this level, they would elect leaders of like kind, who would have no desire for an arms race but only for whatever might serve the greatest good of humanity, starting with the survival of the species.

Einstein's formula was developed into "overkill" in barely 50 years, but Freud's method has been met with constant obstructions and has not been allowed to develop to its potential. First, Carl Jung was branded a "mystic" and his improvements on Freud's method became limited to a small following. Then the insistence on physical evidence made Freud's method look "unscientific." "Freud did bad science," it was said. Freud was demonized and made to look like a bad person. Because Jung's improvements were known only to a small following, Freud's methods had to stand on their own without these improvements and were vulnerable to criticism. Whereas young people such as myself had achieved personal growth through Freudian methods in the early 1960s, a generation that discovered the spiritual in the late 1960s had to go to India to find the answers they were looking for, because they didn't find answers in Western civilization. If Jung's openness to the spiritual had been well known, they might have stayed with psychotherapy for personal growth, but instead they switched to the ancient Eastern disciplines of meditation and yoga.

I fault Western philosophy for not recognizing Freud's discoveries and methods as part of "philosophy." Socrates said, "Know thyself." Freud showed us how to do that.

I define "philosophy" in terms of a question, "How shall I live?" Socrates defined it in a more authoritarian way, "How should we live?" In either case, philosophy is the application of the mind to the living of life. The fact that we sometimes do things that we don't consciously want to do indicates that there must be an unconscious part of the mind. And to the extent that this unconscious determines how we live our lives, it should be included in "philosophy." Recognizing it, dealing with it, bringing it into consciousness, correcting the inaccuracies it creates, and developing the mind to its adult potential are all legitimately part of philosophy (not the philosophy of the ancient Greeks, but the philosophy of our present time), and all of these things are accomplished through psychotherapy.

Given that the human mind is unreliable, we have a method called "science" that helps us to obtain accurate knowledge. But also we have a method called "psychotherapy" which helps to make our knowledge more accurate: Given that the human mind is unreliable, make it reliable. Both of these methods can work together to achieve what I see as the primary goal of philosophy, to maximize the scope and accuracy of our knowledge.

Western philosophy talks about "the good life." Certainly psychotherapy helps people attain "the good life," or what is now known as "quality of life." Western philosophy defines itself as "the pursuit of wisdom." "Wisdom," as I see it, is a long way from the ongoing intellectual argument that is now Western academic philosophy. I see three components to wisdom — accurate thinking, knowledge, and the ability to differentiate. This third attribute is developed in psychotherapy.

The psychotherapy that we, the "Type One Clients," were undergoing in the early 1960s was really philosophy. We were all able to function in the real world. What we were looking for was quality of life and greater wisdom than our academic educations had given us.

And going up the line in human psychological development, compassion and altruism are certainly important factors in how one lives one's life, and a controlled clairvoyance is certainly important in expanding one's capacity for knowledge.

So all those things that psychotherapy offers are legitimate aspects of what philosophy should be. Western philosophy should have assimilated the teachings of Freud and Jung and all the developments in the field of psychotherapy, as important contributions toward the pursuit of wisdom and achieving "the good life."

I think the academic philosophers argue that psychotherapy rightfully belongs in the field of psychology, not philosophy. I would agree that practicing and developing the methods of psychotherapy belongs in psychology. But UNDERGOING PSYCHOTHERAPY belongs in the domain of philosophy. I am shouting this because I think that the psychological defenses of so-called "educated" people prevent them from even thinking that they should undergo a process which they may associate with a lower caste of people who aren't able to function in society. But I hope they will learn, as I have learned, that this development of their adult mental potential is as important a part of their education as the intellectual education which they now think of as the only kind of education there is.

So it seems that my value system, in turning my back on the culture, has some value. It has enabled me to discover things that the culture is not aware of, or actually to rediscover things that the culture has lost, forgotten, ignored, dismissed, confused, or deemed unfashionable, such as philosophy. This is not the philosophy of the ancient Greeks, but a philosophy which is alive and is able to incorporate more recent cultural innovations, such as psychotherapy, which in turn offers a very real solution to a major problem in the living of life, the threat of nuclear annihilation.

The academic philosophers say that if I am working on practical problems, I am not doing "philosophy." Because they have the status, they get to determine what words mean. So I have thought that I should coin another term, such as "software for living," for the application of the mind to the living of life. But no, this is "philosophy" in the oldest and broadest sense, and should be called "philosophy." What the academic philosophers are doing should not be called "philosophy," but should be called something like "the history of Western philosophy." And even though these academic people have the power to say what words mean, those of us who can think for ourselves can understand that only a small fraction of what should be called "philosophy" is actually studied or taught at our American colleges and universities.

We have this wonderful technology, created mainly by science. And at the same time, we have cultural opinions, attitudes, and beliefs that have created weapons that can destroy us, or at least "bomb us back to the Stone Age," as people were saying in the 1960s. The personal development of human beings has not kept up with our technological development. Some of the blame here is on the scientific establishment for holding us to a physical view of the reality. We need to break away from this physical view. Physical science is just fine in dealing with the physical, but physical science should not be allowed to dominate to the point where it obscures our view of the mental and inhibits our mental development.

It has been more than 50 years now since my discoveries of the self-steering process and the psychological age of puberty, and these things are still not known to the culture at large. Part of the problem is the rejection of Freud and the dismissal of Jung. So my whole knowledge base has been undermined. Another part of the problem is the domination of the scientific establishment who dismiss any knowledge that didn't come from the in-group. My personal experience is called "personal," implying that it doesn't apply to anyone else. My evidence is considered "anecdotal evidence," because I don't have scientific credentials. It is considered to have "no referent" because it is based on the evidence of the mental senses. According to the scientific establishment, it doesn't exist.

CHAPTER 14

My Path to the Light

The Theosophical Society talks about putting people on "The Path," as if there were only one path to enlightenment. I tried to tell them that I had found another path, but they didn't pay any attention to me.

Actually, a book I bought from the Theosophical Society, *Beyond Religion*, by David N. Elkins, PhD, described eight different paths, one of which was similar to mine.

My path was through psychotherapy and dream analysis, enhanced by my own discoveries of the self-steering process and the psychological age of puberty, and with major credit to Carl Jung for leading me into the spiritual.

Unlike people who believe in the spiritual just because they believe, I was led to the spiritual through a series of logical steps, like a path, which I can explain to other people, in case they might want to follow that path. If psychotherapy is not limited by the therapist's belief in a purely physical universe, it can lead to spiritual awareness.

I think that what I experienced was roughly the equivalent of Jungian analysis, because it was inspired by the writings of Carl Jung and not limited by any therapist's limitation to the physical. I give Carl Jung the major credit for observing — scientifically, of course, in the face of the scientific in-group and its aura of infallibility — that his clients were experiencing a spiritual reality in their dreams.

My main inspiration came from *The Basic Writings of C.G. Jung*, which led to my winter of dream analysis. This book explains his thinking better than any of his many other books I have. I read the chapters in chronological order so that I could more easily follow the development of his thought.

He doesn't just leap into the "spiritual." He is writing about psychological concepts. He is describing "archetypes," which are symbolic representations of universal human ideas. These archetypes in dreams can be "numinous," meaning that they may have a spooky quality about them. These archetypes were probably the inspiration for the ancient gods and goddesses. Slowly and logically, he is moving from the psychological into the spiritual. He is showing, by observing the content of people's dreams in a scientific way, how our psychology universally includes things that we might call "spiritual."

And then he tells the story of the monk who prayed that he might look upon the face of God and was granted his wish. The experience was so overpowering that it made him insane for 15 years until he finally assimilated it psychologically. Jung called this "original experience."

For me, as a child of the mid-20th century, God was an entity thought to have been seen only in ancient times and only by a spiritual elite. It had never before occurred to me that it was possible to see God for oneself. Jung brought this idea into my consciousness with the concept of "original experience." This, in my opinion, was Jung's greatest discovery, that it was not necessary to believe in the spiritual on the basis of "faith," but possible to believe on the basis of EVIDENCE.

I have to shout that word "EVIDENCE" to wake up everybody, because after 1850 years of brainwashing it is only thinkable that the spiritual must be believed on the basis of "faith." Actually the Catholic Church, in the 2nd century AD, decreed that everybody had to believe on "faith," and killed off the Gnostics, who believed on the basis of evidence.

The evidence, as I experienced it, was the evidence of the mental senses, obtained through the mechanism of dreams. And where this might be dismissed as subjective and therefore unreliable, the self-steering process makes it reliable, actually more reliable than the evidence of the physical senses, because the physical senses have no self-steering process. I have accepted everything I am about to say as part of my belief system, and my dreams have never corrected it. And where my dream experience would not be exactly replicable in other individuals, because we are all coming from different places, I would expect our dream experiences to lead us all in the same general direction.

From my dreams I encountered the greatest wisdom I had ever experienced. My dreams helped me to resolve my most serious traumatic experience, which my psychiatrist had missed. They also helped me grow to the psychological age of puberty. This same process that led me to psychological wellness also introduced me to the spiritual. Thus I was led to believe that the spiritual was equally real and equally beneficial to my life.

The spiritual was absolutely necessary for me to resolve the traumatic experience. Briefly, an angry adult male had terrified me at age two and a half, and in my childhood misunderstanding had caused me to draw many wrong conclusions which had crippled me mentally for 30 years. When I recognized the situation, I thought, "I am 30 and he is probably now 70, and I can find him and beat the shit out of him." This had been my childish fantasy of revenge for 30 years. But with the influence of the spiritual, I felt compassion for him. He was just a poor guy with problems of his own, venting his anger on a small child. I forgave him for his mental injury, and that resolved the traumatic experience.

Compassion was also necessary for me to reach the psychological age of puberty. I had to feel love for all humanity, a love for the nameless "man, woman, and child" that was about as strong as my love for my own survival, in order to risk my life by stepping into that little boat, not to be the hero, but only to play some small part in their rescue.

In order to teach me compassion, the dreams spent a long time teaching me the meaning of "love." The Beatles sang, "All you need is love," but what, really, is love? Love is something you fall in. Love is desire. Love is self-gratification. Love is many things. Everybody wants it. People give it to somebody who can do something for them. People told me they loved me in order to manipulate me. To help me go from the selfish kind of love to the unselfish kind of love, or compassion, the dreams gave me many lessons.

They started off with combat lessons, as appropriate for a 10-year-old boy, defending himself from (or being killed by) wild animals, such as reptiles, bears, lions, and wolves. Gradually from this emerged another theme, one of cooperation rather than competition. There were lessons of trust and friendship between human beings — instead of the usual business contract, more of a trust that money would be paid, as in giving somebody a tip.

From these simple lessons of trust and friendship, the dreams went on to show how love could overcome my fear of animals that could kill me. I was rubbing against a lion in the dark, thinking it could kill me, and then thinking of its feelings and saying, "Oh, you poor thing, all alone in the dark." Then the sun came out and there was no reason to fear.

Light, in my dreams, was the symbol of love, from turning on a small light to turning on the sun.

And then there was the supreme Light. I dreamed that behind a fog bank there was a very powerful Light, many times more powerful than the sun. And in front of the fog bank there was a small gold cross. There was no doubt in my mind that the powerful Light was God, and the small gold cross was Christ, "who makes intercession for us." And the supreme Light was the supreme Love, as I had seen on the wall of the Christian Science Church, "GOD IS LOVE."

I have since then stopped calling it "God," because I don't know that this power is an entity with a will and all the attributes given to it by the Christian Church, but I do know that this supreme Light is the supreme Love, which is the supreme power of the universe, at least in our human relationships.

The spiritual was first represented in my dreams by a ship that steered an "S" course, for "soul" or "spirit." There was no question about that, in my mind. The ship was going out of Provincetown harbor on a course that would take it into the harbor. There was a cold wind blowing from the south, not north. I learned to recognize the spiritual in dreams because it was the reverse, or mirror image of the physical. The waterfront of Provincetown faced north, instead of south, in a spiritual dream.

I can hear the intellectuals saying, "'S' can stand for many things. Dream interpretation is ambiguous. It is like reading tea leaves."

Dream interpretation is not intellectual; it is intuitive. The interpretation of the dream must be consistent with all the details of the dream. The dream is exactly what the dreamer needs to know in the dreamer's particular life's situation. That narrows down the possibilities somewhat.

And if your interpretation is really wrong, there is the self-steering process to correct you. And none of this, for me, has ever been corrected.

The spiritual, of course, was the thing that was seriously missing when I said goodbye to my psychiatrist. I needed to recognize the spiritual and assimilate it into my being, in order to feel like a whole human being. The wooden puppet, Pinocchio, needed the spiritual in order to become a real person. Whether the spiritual is real or not, it seems that this is a universal human need, because every culture on earth has some kind of spiritual belief.

I am sure that the physicalists will ridicule me at this point by saying that I had become a "believer," but actually the acceptance of the spiritual in my life was a major component in reaching my goal of happiness.

So instead of dismissing Jung as a "mystic" and me as a "believer," scholars need to examine all this in a scholarly way: Is this experience of the spiritual only some dream fantasy, with no relation to the real world? Is this experience of the spiritual only a force within the human being, or does it represent a Force in the external world? Does the fact that every culture in the world has a belief in the spiritual have its origins in this universal dream experience? How may this dream experience be interpreted, other than in the conventional religious way?

When I had resolved the problems of my childhood, then, I wondered, what would I dream about? I was having some really strange dreams that didn't make any sense, until finally this dream gave me a clue:

Movie camera car, taking pictures, coming to a railroad crossing, almost colliding with huge locomotive coming on tracks at an oblique angle. Car doesn't stop, just keeps going at a constant speed, approaching intersection as light turns yellow, red, then instantly green as camera car goes through. I remember having seen this before, look in rearview mirror of camera car and see my '48 Olds following behind. I am trying to get a better look to see if it is really me.

My '48 Olds represents the past. So if I am looking in the rearview mirror and seeing the past, then of course what I am seeing through the windshield is the future. This camera car which travels at a constant speed represents the passage of time. The camera represents my ability to see ahead of me in time as well as behind me in time, not forever, but in some limited range, as a camera would see.

This explained some of my confusing dreams. As I reviewed them, I could see that many of them were about future events.

I dreamed that a friend came to see me, a friend I had not seen in about a year. He showed up the next day. I told him about the dream. He said that it could have been caused by telepathy — that he was planning the trip, and he knew he would see me, and I could have picked up his thoughts. I couldn't argue with that. But in December of the following year, when I was living on Beacon Hill in Boston, I had the following dream/experience:

It is a very cold morning as I start off to work. My car is parked against the left-hand curb, facing downhill. Thus the door swung open downhill, and I left it open as I got in to start the car. I turned the key and nothing happened — it wouldn't start.

The dream was at about 7 AM, and at about 8 AM I started off to work. It was a very cold morning. My car was parked against the left-hand curb, facing downhill. Thus the door swung open downhill, and I left it open as I got in to start the car. I put the key in the ignition, and at that point I remembered the dream. I had never had any trouble starting the car in the five years I had owned it, so in total confidence that it would start, I turned the key. And nothing happened — it wouldn't start.

There was a subtle problem in the starter wiring, which was to cause me a great deal of trouble and expense. I made 5 trips to 3 different garages before an expert mechanic finally was able to fix the problem.

First of all, let's get rid of the "coincidence" argument: The car had started perfectly for 5 years and 9 months, or about 2000 days, and the dream had prophesied exactly the time that it first failed to start. That's one chance in 2000 of it being coincidence.

And then, to get rid of the "telepathy" argument, here was a machine that had a malfunction that nobody knew about, until the dream revealed it to me.

This shift to dreaming of the future is consistent with the teachings of Richard Kieninger, that when one has reached psychological adulthood, the next step in psychological development is to develop one's clairvoyant powers. It is also consistent with the idea of spiritual bypassing, in that when one has resolved one's childhood traumas, not only is one free to develop spiritually, but also one's dreams are supporting the person in that forward development, instead of always looking backward at unresolved problems.

I had seen evidence of a spiritual reality, but I knew very little about it. I had studied physics to learn about the physical reality, and psychology to learn about the mental reality. Now I needed some kind of education to help me understand the spiritual reality. Just as I was pondering this, an old girlfriend came to town and handed me a copy of *There Is a River: The Story of Edgar Cayce*, by Thomas Sugrue.

Briefly, for those who aren't familiar with him, Edgar Cayce (pronounced "Casey"), American psychic, 1877-1945, was able to go into a trance state, and in that trance state was able to answer questions about things that he did not consciously know, and other people couldn't see, such as people's medical problems, or the Akashic Record, which he described as "God's book of remembrances." These trance sessions were called "readings," and present at these readings were always a "conductor," who asked the questions, and a stenographer, who recorded everything that was said. The stenographer, of course, was very important, because now we have a permanent record of all these readings. These readings are now on file at the library of the Association for Research and Enlightenment (A.R.E.) in Virginia Beach, and are available for qualified researchers to study.

Most of these readings were medical diagnoses and suggestions for treatment. Cayce always worked with a licensed medical doctor, who made the actual prescriptions. Somebody followed up, or tried to, on all the treatments, to see whether they were effective. That information was documented, too. And in more than 90% of the cases, his treatments were proven to be effective. The other 10% weren't all proven to be ineffective, because some people didn't follow the treatments, and some didn't report back. This rate of success is even more remarkable because a great many of these were cases where conventional medicine had failed.

After Cayce had been doing these medical readings for a while, a man named Arthur Lammers came and asked for a horoscope in a reading. Without having been prompted, Cayce volunteered the information that this was the third incarnation on earth for Lammers. This was the first of about 2500 "life readings" that Cayce gave, describing people's previous incarnations and how they influenced their present lives.

For me, this was the proof of reincarnation. Here was an unknown mental process that produced verified accurate results in more than 90% of 9000 medical readings. So this process had a verified error rate of less than 10%. Then what was the probability that the process was wrong in all 2500 life readings, which could not be verified, that asserted that reincarnation was a fact? Note that this is not the probability of any one particular life reading being accurate, but the probability that every life reading was accurate, in asserting that there was such a thing as reincarnation, and never denied. Given an error rate of 10%, the probability of Cayce having been wrong about reincarnation in every one of 2500 life readings is $.10^{2500}$, a number so small that it is meaningless. The probability of reincarnation being a fact, based on this evidence, is as close to certainty as you can get.

Was there any bias operating here, that would have biased Cayce towards a belief in reincarnation? First of all, no suggestion of "reincarnation" had been given to him in the initial life reading. He had been asked for a "horoscope," and he spontaneously came out with the information on reincarnation.

What makes it even more convincing is that his bias was against reincarnation. He was a devout Christian, who read the Bible once for every year of his life, and the Bible didn't say anything about reincarnation. He resolved this conflict by observing that the Bible didn't say anything against reincarnation, either. He even found passages in the Bible that hinted at reincarnation:

John 8:58 ... Before Abraham was, I am.

John 9:2 ... Who did sin, this man, or his parents, that he was born blind?

And he was in constant prayer: "If this isn't right, I don't want to do it."

To support Edgar Cayce's assertions that reincarnation did exist, *Twenty Cases Suggestive of Reincarnation* by Ian Stevenson has now been followed by evidence of reincarnation in more than 2000 such studies. Some of these have been in the United States, where people don't even believe in reincarnation. There is now a six-step procedure that reincarnation researchers follow, to protect against bias and cheating.

Along with reincarnation comes the Law of Karma. This is the other half of the Golden Rule, "Do unto others as you would have them do unto you," because whatever you do unto others will be done unto you — maybe not in this incarnation, but in some incarnation. "Those who live by the sword shall die by the sword," again, maybe not in this lifetime but in some lifetime.

You don't have to believe in reincarnation to see the Law of Karma in action. The force that makes people keep recreating those situations in life where they have psychological problems is at least related to the Law of Karma: "You have to keep coming back until you get it right," as one of The Beatles once said.

Reincarnation and karma are only a couple of the subjects dealt with in the Edgar Cayce readings. Other subjects include Atlantis, levitation, a cosmology, and the idea that the purpose of our life on earth is education towards our spiritual advancement. These ideas represent an enormous contribution to our culture.

But Edgar Cayce, it is insinuated, it is whispered, in academic circles, was somehow fraudulent. His critics need to produce evidence supporting this claim. I have seen no evidence of fraud in the dozens of books I have read about Edgar Cayce or the many dealings that I have had with the A.R.E. over the years.

Edgar Cayce had to go to court twice, once on charges of practicing medicine without a license, and another time for "fortune telling." He was acquitted both times.

Two scholarly investigators were reported in *There Is a River*. The first was Hugo Munsterberg of Harvard in 1912, who concluded that the phenomenon warranted further study. The second was William Moseley Brown of Washington and Lee in 1928, who said, "I can't expose it. ... I'll have to believe in it."

Everything I know about Edgar Cayce appears to be totally genuine. If academic people have evidence to the contrary, they need to come forward with it. Their beliefs seem to be more in line with the insinuations I have read in skepdic.com, which their freshmen students in critical thinking should be able easily to invalidate.

My dreams warned me of the dangers of taking LSD, but I had to experiment with marijuana in the summer of 1967, just to catch the flavor of the drug experience. Some of it was good, but some of it was bad. My natural defenses were opened up so that dead people were coming into my dreams, and this was frightening to me.

In 1968, I brought this problem to a highly recommended psychiatrist in the Boston area. His opinion was, "You have a deep-seated problem." I told him that I had already successfully completed psychotherapy with another psychiatrist, and that I had enough experience with dream analysis so that I could tell the difference between what was coming up from my subconscious and what was coming in from an outside entity. But his belief system was fixed. He didn't believe in the existence of spirit entities, so anything to do with "the dead" had to be a deep-seated problem.

I even wrote to my previous psychiatrist, asking him to verify that I had already successfully completed psychotherapy. But I realized that he didn't believe in spirit entities, either. I never heard from him.

This brings up the legal issue that two psychiatrists sharing a fixed belief system can put away for life, without due process of law, a person who doesn't agree with their beliefs. There is the assumption that the psychiatrists are right, because they have credentials — the caste system — and the dissenting person is wrong. There is a certain arrogance in this, in assuming that the people with the credentials know all there is to know.

I asked myself, "Why am I paying this psychiatrist top dollar to give HIM an education?" So I ended the treatment and sent him a copy of *Edgar Cayce on Dreams*, which I'm sure he immediately threw into the trash.

I was beyond the scope of the knowledge of the psychological profession, beyond the reach of the culture to help me. I was on my own to deal with my psychological problems.

Edgar Cayce had taught me that we, the living, were more powerful than spirit entities, because we had everything they had, plus a physical body. That helped me, initially, to get rid of the spooks in my dreams. I yelled in one dream, "Get the Hell out of here!" and they all vanished. A simple assertion of will was enough. Later on, I learned The Protective Prayer from Richard Kieninger: "Dear Christ, please protect me." This protection from higher beings was almost miraculous in bringing me up out of a spooky dream instantly.

Yes, there is such a thing as spirit interference, as people believed before the scientific establishment decreed that there was no such thing. But the individual is not morally responsible for these "demonic" influences, any more than children are responsible for being attacked by the bullies in the schoolyard. And while the psychiatric profession is lagging in its knowledge, we can ask for the protection of higher beings, also believed not to exist.

When I reached the psychological age of puberty and became an altruistic person, it seemed that there was no place for me in the world, in a culture dominated by competitive self-interest. I had read somewhere that it was possible for people to will themselves to die, and I was afraid I might will myself to die. The dead people in my dreams enhanced that fear, as if they were expecting my arrival.

I lived in that state for about two years, until the spring of 1969, when I was organizing the material for *Re-Educating Myself.* One day I was exhausted and depressed, seeing the huge amount of material that I had and the enormous task that lay ahead of me. I flopped down on my bed, not caring whether I lived or died.

Immediately I lost physical consciousness. Immediately I shot up out of my basement apartment in a tunnel just wide enough for my shoulders, up, up, up, an enormous distance at an enormous rate of speed, into the presence of the Light. I knew from reading Carl Jung not to look upon the face of God, so I shielded my eyes from the Light with my right arm. Off to the left, or at the right hand of God, stood the figure of Christ. He told me simply, "Get back to work," and back down I went.

I have never again been afraid to die or afraid that I might will myself to die. I have felt since then that the higher powers were supporting my work — that I have been on "God's payroll."

Not long after that, I was walking the dunes on a beautiful sunny day thinking beautiful thoughts, and when I came into town, I saw a beautiful woman with a friend of mine, so of course I had to meet her.

"You are very disturbed," she said.

I tried to explain to her that it was a beautiful day and I was thinking beautiful thoughts and I was definitely not disturbed. But she kept insisting that I was disturbed. And the more I tried to argue with her, the more disturbed I became. When I finally managed to break away from her, the sky had clouded over and it wasn't a beautiful day any more.

That experience was my introduction to the Forces of Evil (FOE). As there were positive spiritual forces in the universe, so also there were negative spiritual forces. As I had begun to think of myself as being on "God's payroll," now here I had encountered somebody working for the opposing forces.

I am sure that this one piece of evidence that there is such a thing as the Forces of Evil (FOE) will be met with ridicule and scorn by the physicalists. But please note that ridicule and scorn are the only arguments they have ever had to dismiss the idea of such forces existing. There is no scientific argument, and ridicule and scorn actually prevent science from operating, because they create a bias against even investigating the subject.

All of this, I realize, is personal and unprovable to another person. But most of life is personal and not provable to another person. Where science can't always be applied in its authoritarian form, scientific methods can be applied in our daily lives to prove things to ourselves. This I believe is the meaning intended by Richard Kieninger when he describes the Brotherhoods as "scientist/philosophers." We can all use the methods of science, of observation and proof, in our daily lives, to determine what is more accurate and less accurate, approaching an ultimate knowledge of what is true and not true, without the need for so-called "authorities" to tell us what to believe. And so I introduce

"The Brotherhoods' Philosophy," as best as I am able to represent it, as presented by Richard Kieninger, also writing under the pen name of Eklal Kueshana.

The Brotherhoods are secret societies of people who are highly developed mentally and spiritually. They keep their identity secret to avoid persecution. They recognize each other through controlled clairvoyance. I had heard about the Masons, who incorporated their own spiritual symbols into the building of the cathedrals of Europe, and the Sufis, whose poetry incorporated a kind of code of spiritual symbols, so I was ready to accept the idea that such people existed. These are living people on earth, not spirit entities. The information coming through Richard was not "channeled."

The first thing anybody needs to know about the Brotherhoods' philosophy is that they were wrong about the Doomsday that was supposed to occur on May 5, 2000. That should get rid of the people who need to believe on the basis of somebody else's so-called "authority." The Brotherhoods' philosophy is something to believe on the authority of your own mind. You have the authority and the responsibility to decide for yourself whether or not they are right.

And to start with, as I have already mentioned, it should all be taken as a working hypothesis. Yes, they offer a whole cosmology, one that makes more sense than the Christian teachings, but this is not to be believed absolutely as Gospel, but only as a hypothesis for each of us to test in our life experiences and prove or disprove for ourselves.

Consistent with this approach, the Brotherhoods teach that what you read in a book is not "knowledge" but only "information." It doesn't become "knowledge" until you have tested it and proved it to yourself in your life's experience. This I also agree with, on my own authority, as the highest authority for deciding what to believe for myself.

One of the things I like about the Brotherhoods' philosophy is The Twelve Great Virtues: sincerity (honesty), courage, devotion, charity, patience, humility, precision, efficiency, kindness, tolerance, forbearance, and discernment. They are things I have been working on all my life in my psychological development, without actually naming them. In fact you can't work with them as just an external chart on the wall that you check off every day.

You have to look inside yourself and ask, "How was I honest, courageous, kind, tolerant, and so forth today." They go hand in hand with psychotherapy.

And I also appreciate the fact that some things that are considered virtues in other systems, such as obedience, chastity, and vegetarianism, AREN'T there. These are NOT virtues, to my way of thinking, again on my supreme authority to decide for myself.

Because the word "virtue" comes from the Latin "vir," for "man," I would change the name to something else, such as "The Twelve Great Human Qualities," to get away from the masculine terminology.

The most powerful and the most frightening, to me, of the Brotherhoods' teaching is the idea that there are spirit entities called "Black Mentalists," whose karma is so bad that they cannot possibly incarnate on this earth, because they would self-destruct immediately. So they can never develop spiritually and are doomed forever to weeping and wailing and gnashing of teeth in utter darkness. Therefore they attempt to bring everybody else down with them, by polluting our minds with telepathic suggestions. No living person on earth is a Black Mentalist, but living persons can be influenced by Black Mentalists, as can be detected by black in their auras. We can see with our normal senses the influence of the Black Mentalists in the black limousines and suits of the very rich and powerful and the black robes of the priesthood.

The first two people I told this to instantly called me "paranoid." That sounds like an instant psychological defense against a frightening idea. And it is possible that Black Mentalists helped out in that defense, because they put ideas in people's heads. And possibly all psychological defenses are influenced by Black Mentalists, who don't want anybody to have a clear head.

We can see the evil that exists in the physical realm. So I think that New Age explorers into the spiritual are naive when they make no mention of possible evil in the spiritual realm.

Again, this is not Gospel; this is a working hypothesis. Might there be evil entities influencing our religions, our politics, and even our most highly respected sources of knowledge, our aca-

demic institutions, by malevolent telepathic suggestions? I leave
that as a question, but it is not a question to be ignored.

Meanwhile, according to the Brotherhoods, the Holy Spirit
protects us against Black Mentalists. The Holy Spirit is composed
of the most highly developed human beings, along with angels
and archangels and higher celestial beings. All of these entities are
more powerful than Black Mentalists. We have all these higher
beings to protect us, and the Protective Prayer, which is basically,
"Dear Christ, please protect me."

Richard Kieninger has taught that we should not pray indis-
criminately to the spirit world for "help" or "guidance," because
there are plenty of spirit entities ready to guide us to our doom,
but that we should pray specifically to Christ and only for protection. He was accused of religious prejudice when he said, "Pray
specifically to Christ," so he said, yes, of course you can pray to
Buddha or Mohammed or Moses or other high-level Adepts.
They are all members of the Holy Spirit and all more powerful
than the Black Mentalists.

I don't believe in angels, but I pray to them. It works.

As with Edgar Cayce, there are many aspects of the Brotherhoods' philosophy as presented by Richard Kieninger that I haven't mentioned here. And as with Edgar Cayce, these are ideas
that need to be incorporated into our culture.

At the dream conference in 1991, I was having lunch with a man
in a Hawaiian shirt. He asked me, "Have you ever considered
that dreams might come from God?"

I hadn't thought about that. Dreams came from the subconscious. But "subconscious" is another one of those words like
"random" or "placebo effect" that we use when we don't know.
We don't know where dreams come from. Looking for dreams in
brain activity might be as inappropriate as trying to understand
TV programming by studying the workings of a TV set. Dreams
may come from God, for all we know.

The man in the Hawaiian shirt later identified himself as a
Catholic priest. If he had been in uniform, I probably would have
dismissed his question as simply a reflection of his belief system,
but because he wasn't, he gave me the opportunity to think about
it. These dreams, which were the greatest wisdom I have ever

encountered, may not have come just from my own subconscious, but may have come from higher beings.

This is my path to the Light. It is not complete. I don't know what is at the top, but I know that if I keep going uphill I will get there.

My approach is through psychotherapy, through self-knowledge, and through self-correction, with help from my dreams. Dream analysis, taking advantage of the self-steering process, is a natural path to spiritual growth and enlightenment. Dreams are tailored precisely to the educational needs of the individual, as opposed to man-made disciplines which may demand some conformity to a philosophy and take the individual away from his/her real self.

I have heard that people on "The Path" are required to submit to, and be obedient to, the absolute authority of the Teacher. On my path, I don't need any "authority" telling me "The Truth" and demanding my absolute obedience. The psychotherapist is only a guide, to help me find the places where I know subconsciously that I am wrong, on my own authority. Dreams are the same. They may appear to be an "authority," because of their ability to correct me, but it is all subject to my interpretation. I am the authority who has to agree with what the dreams are trying to say.

My ability to communicate this path to the culture has been blocked at every step of the way by academic prejudices against Freud (not during my therapy, but more recently), psychotherapy (as philosophical education), Jung (always branded a "mystic"), dream analysis ("Dreams are a random firing of neurons"), the mental senses ("not objective"), spiritual reality ("nonexistent"), Edgar Cayce (insinuations and laughter), and Eklal Kueshana (whose book was sold UNDER THE COUNTER at a Boston bookstore). I challenge the academic establishment to defend each one of these prejudices with an argument that would receive a passing grade in their freshman courses in critical thinking.

CHAPTER 15

J. B. Rhine and Parapsychology

I first heard of J.B. Rhine and "extra-sensory perception" (ESP) when I was at Harvard in the 1950s. The subject wasn't taught in any of my classes, but it was generally presented as a reputable and legitimate study, probably because it was done at Duke University.

I don't use the expression "extra-sensory perception," because I feel that it is a contradiction in terms. I define "senses" to mean any means by which we perceive anything. J.B. Rhine used the expression "ESP" to convey that there are mechanisms of perception beyond the physical senses recognized by the culture. And of course they are also beyond the mental senses that I have recognized. Some people have called them "higher senses." More recently, people in the field of parapsychology have used the word "psi" to represent these abilities or phenomena.

The field of parapsychology includes the study of telepathy, which is the ability to read other people's minds, clairvoyance, which is the ability to sense things that aren't in anybody's minds, precognition, which is the ability to see the future, and psychokinesis (PK), which is the ability to affect physical objects with the mind. The psi concept has also been extended to include energy healing (and its opposite, voodoo), a form of PK, the power of prayer, another form of PK, and communications with spirit entities, a form of telepathy. Parapsychology also includes postmortem survival and reincarnation research.

Unlike the early psychologists who were working largely with mental evidence, J.B. Rhine managed to conduct his experiments by the strictest rules of physical science. So there should have been no complaints from the physical scientists. But still there were complaints, starting with the absolute assertion that such things as psychic abilities were impossible, and that therefore he must have cheated, or at least cued his subjects in some way.

The physical scientists have taken great delight in something that has been called "the Rhine effect." Subjects who do very well to start with, in tests of psychic abilities, tend to do worse as time goes on. Sometimes this is after more than 1000 trials. I think they just get bored.

Another explanation for "the Rhine effect" might be that mischievous spirits give them the psychic ability in the first place, and then go away to play elsewhere.

Parapsychology has had official status as a "science" since 1969. In that year the application of the Parapsychological Association (PA) for affiliation with the American Association for the Advancement of Science (AAAS) was officially accepted, with help from this statement by Margaret Mead:

> For the last ten years, we have been arguing about what constitutes science and scientific method and what societies use it. We even changed the By-Laws about it. The PA uses statistics and blinds, placebos, double blinds and other standard devices. The whole history of scientific advance is full of scientists investigating phenomena that the establishment did not believe were there. I submit we vote in favor of this Association's work. (quoted in Broughton, 1991, page 74)

> (Quoted in E. Douglas Dean, "20th Anniversary of the PA and the AAAS, Part I: 1963-1969." *ASPR Newsletter*, Winter 1990, pages 7-8.)

Richard Broughton goes on to say, "Following her statement, the membership voted five-to-one in favor of granting that affiliation."

It is generally known that experiments in parapsychology are conducted more rigorously and more carefully, on the average, than experiments in other scientific disciplines, because the subject is met with so much suspicion. And yet, in 1979, Broughton says, there was a campaign to "drive the pseudos out of the workplace," which failed.

Parapsychology had to defend itself against this attack and reinstate its good standing in the AAAS. I would have turned that around. I would have identified the person doing the complaining as a zealot using smear tactics in the name of science. I would have pointed the finger at him as the one corrupting the workplace. I would have suggested that HE didn't belong in the AAAS.

After that incident, parapsychology has maintained its good standing in the AAAS as a legitimate science. And yet when I looked up "parapsychology" in Wikipedia in January 2016 (it can change at any time), it started off by saying, "Parapsychology is a pseudoscience ..."

The person best known for conveying the message that parapsychology is a pseudoscience is magician James Randi. He has offered $10,000 to anybody who can successfully demonstrate psychic abilities, increasing that to $1,000,000 in recent years, and nobody has been able to collect it. That's because Randi makes the rules and has set himself up as the judge to decide who gets the money, and he is absolutely dedicated to disputing every single instance of psychic abilities. And he doesn't limit himself to scientific arguments. We can all read his bias in his book title, *FLIM-FLAM!: Psychics, ESP, Unicorns, and other Delusions.* If there is any doubt about his bias, we can read the following smear words just on page 1: "pandering shamelessly," "rickety bandwagon," "feeble rationalizations," "hoaxes," "shrinking public," "purported wonders," "quackery," "scientific tragedies," "self-deluded," and "psychological trickery." And yet this man is cited more often in contemporary psychology textbooks than J.B. Rhine (McClenon et al., 2003).

CHAPTER 16

Freshman Orientation

I was surprised to find a highly biased propaganda piece against parapsychology in the textbook for the beginning psychology course at Dartmouth College in the fall of 1998 (Bernstein et al., 1997), illustrating how social signals are sent to young and impressionable college freshmen, insinuating that parapsychology is somehow disreputable. I'll quote just a few choice bits here. I have capitalized certain words to point out the bias that is being applied.

> ... Throughout history, however, various people have CLAIMED *extrasensory perception*, the ability to perceive stimuli from the past, present, or future through a mechanism beyond vision, hearing, touch, taste, and smell. ALLEGED forms of extrasensory perception (ESP) have included *clairvoyance* ... , *telepathy* ... , and *psychokinesis* ... . CLAIMS for ESP are widely BELIEVED (Alcock, 1995; Lett, 1992). Research on them is called *parapsychology* ... . (Bernstein et al., 1997, pages 131-132)

The words that I have capitalized convey the desired bias — that this is just somebody's belief system and not to be taken seriously.

The references above to Alcock and Lett are to *Skeptical Inquirer*, which I like to call "Skeptical Enquirer" because of the high level of inaccuracy of its methods and its content. I am not including these people in "academia," not wanting to pollute academia by including them in with reputable academic people, but I see here that the people in academia don't mind polluting themselves.

> Many APPARENT parapsychological phenomena are weak and difficult to
> replicate. Under close scrutiny by OUTSIDE OBSERVERS, reported ESP
> phenomena often FAIL to occur [Druckman and Swets, 1988]. (Bernstein
> et al., 1997, page 132)

Again, I have capitalized words to show the bias being con-
veyed. It is true that parapsychological phenomena are weak and
parapsychological tests often fail. I have capitalized "outside observers" because this carries with it the bias that there are "inside
observers" who are biased and "outside observers" who are unbiased. This insinuation is not accurate.

Actually there are people who believe in parapsychology and
people who don't. Parapsychological abilities are so delicate, so
sensitive, that an experimenter who is biased against parapsychology can actually block the perceptions through his/her mind-
jamming abilities or just creating a general aura of negativity. It is
like asking the subject to sing a tune while another tune is being
played loudly in the room, or having a wine-tasting session in a
kitchen where sauerkraut is being cooked.

I have actually seen results of a parapsychological experiment
where the subjects scored significantly LESS than chance, and
the experimenter concluded that since the results were "negative," parapsychology was disproved.

If you have ever studied mathematical statistics, you know
that that conclusion is wrong. The result is significant, therefore
there was a force operating to cause the person to score less than
chance — most likely a negative bias on the part of the experimenter and/or the subject. Either or both were probably using
their parapsychological abilities to make the results come out to
support their bias.

So let us not assume that these "outside observers" are either
unbiased or exert a neutral effect on the experiment. Some people, because of their strong negative bias, are simply not qualified
to perform experiments in psychology. All it takes is a little smile
at the beginning: "Today, we are going to test your psychic powers," (smile).

What the authors neglect to mention here, in order to convey
their bias, is that the *Journal of Parapsychology* has been reporting
successful experiments in parapsychology since 1937, and the
British *Journal of the Society for Psychical Research* has been reporting
successful results for much longer than that. To counter the

claim that parapsychological tests "often" fail, there is this vast body of research to show that experiments in parapsychology have more often been successful.

It is interesting that the authors here don't use the word "psi" anywhere. Maybe they aren't aware of it. But they refer to "ESP" over and over again. I don't know their intentions, of course, in doing this, but I do know that the effect of using this expression, which is a contradiction in terms, is to convey the message at some subconscious level, "Perception without a sense is impossible; therefore ESP is impossible."

The text continues:

> In one set of experiments, subjects were asked to predict which of four randomly illuminated lights would appear next (Schmidt, 1969). After several thousand trials, a few subjects correctly predicted the illumination up to 26.3 percent of the time, a performance that — statistically speaking — is significantly better than the 25 percent to be expected by chance alone (Rao & Palmer, 1987). (Bernstein et al., 1997, page 132)

They seem to have picked the least impressive result of a parapsychological experiment they could find. Mathematically, after several thousand trials, the result may be significant, but intuitively, by Weber's Law (which they have just been discussing), a person perceives that 26.3 percent and 25 percent are essentially the same.

They didn't mention Hubert Pearce, who scored 39.6% in 300 trials, where he would have scored 20% by chance. In 1850 trials he scored only 30% (maybe he got bored), but still the odds against doing this by chance are 22 billion to one (Broughton, 1991, page 69).

Continuing with the text:

> Many psychologists have also challenged the use of apparent changes in random number - machine output as a measure of psychokinesis. They note that randomness is not an all-or-none concept. If you flip a coin seven times and it comes up heads each time, is the pattern really nonrandom? Does it mean you have psychokinetic ability? A sequence of seven heads in a row is unlikely (it will occur on average only once in 128 tries), but it certainly can happen. Indeed, a seven-heads sequence is just as likely as any of the 127 other possible seven-flip sequences. In other words, it is far more difficult than one might think to determine that something really is "nonrandom" (Hansel, 1980). Thus, a very plausible interpretation of apparently

nonrandom events in ESP experiments is that they are random after all. (Bernstein et al., 1997, page 132)

This is just wonderful mathematical mind-jive, aimed at creating doubts in the minds of young freshmen who may not have studied mathematical statistics, just as a lawyer might try to create a "reasonable doubt" in the minds of jurors. But science operates with a "reasonable doubt," and there are accepted standards of non-randomness. If there is less than one chance in 20 that the outcome of an experiment could have occurred by chance, the results are considered "significant." If you are anxious that the one chance in 20 might actually have occurred by chance, you can do the experiment again. If you get the one chance in 20 again, then the probability that the results for both experiments combined are due to chance is one in 400, and so on. Generally accepted confidence limits for psychological experiments are two standard-deviations (one chance in 20) and four standard-deviations (one chance in 10,000) of arriving at the outcome by chance. What the authors don't seem to realize is that their mathematical mind-jive, if accepted, would invalidate a great many scientific experiments, not only in parapsychology, but in psychology and other fields as well.

A photograph of a woman in a bizarre position calls our attention to the box in the wide margin of the textbook. The caption reads:

AN ESP EXPERIMENT   This woman is attempting to use clairvoyance to "see" objects in another room while hearing only white noise and wearing goggles that provide unpatterned visual stimulation. She is unlikely to succeed. (Bernstein et al., 1997, page 132)

The woman is made to look foolish, and thus parapsychology is made to look foolish. This is the unscientific method of ridicule.

The evidence (ganzfeld procedure) indicates that she IS likely to succeed, despite the authoritarian pronouncement here, "She is unlikely to succeed."

The caption continues:

... Since 1964 James Randi, an expert magician and ESP skeptic, has carried a $10,000 check that he will give to anyone who can perform clairvoyance, telepathy, or any other feat of ESP under scientific conditions (Randi, [1982]). After investigating hundreds of claims, he still has his money. (Bernstein et al., 1997, page 132)

That's because James Randi makes the rules and has made himself the judge. I have suggested to The Rhine Research Center that they consult with a good law firm to see if they could collect the money from Randi in an impartial court of law. Also they might collect a few more dollars for the defamation inherent in his book title, *Flim-Flam!*

Finally, some apparently remarkable results of ESP experiments can be attributed to fraud. A few researchers have tampered with their equipment and measurements, thus destroying their credibility and raising suspicions about all parapsychological research. (Bernstein et al., 1997, page 132)

This is a nice propaganda move. Now that our brains are befuddled by the mathematical mind-jive and insinuations of "flim-flam," the authors are trying to slip this accusation of "fraud" past our critical faculties for the killer blow. Note that they can say the word "fraud" out loud, since it is aimed at nobody in particular, but also note that it applies to nobody in particular, unless the person is actually identified. It is just a broad-brush smear, as in racial prejudice or class prejudice, to taint the reputation of the whole class of people it is aimed at.

Accusations of "fraud" are not part of the scientific method and are not necessary in science. The key to science is replication. If a finding is fraudulent, it will simply not be replicated by reputable investigators, and therefore not accepted as scientifically demonstrated. It is as simple as that.

There are instances of fraud in every branch of science. So by the same faulty reasoning, the words "thus destroying their credibility and raising suspicions" can be applied to all scientific research. Again, the smear tactics aimed at parapsychology can be applied to other branches of science as well. The propaganda backfires.

Here they conclude:

As a science, psychology depends for evidence on clear, reliable, and replicated observations. Skeptical scientists place the burden of proof for the existence of ESP on parapsychologists, who must be able to show robust, replicable ESP effects that do not depend on interpretations of nonrandomness and that occur under tightly controlled conditions. ... (Bernstein et al., 1997, page 132)

Given the tactics they used, we would expect that the authors would arrive at these conclusions, or actually that they had begun with these conclusions. To counter these insinuations, first of all, scientists have recognized parapsychology as a science by granting the Parapsychological Association affiliation with the AAAS. These authors make no mention of that. And then I offer the following quotes from people who are actually expert in the field of parapsychology, for the freshman student to decide which viewpoint sounds more accurate:

Since the publication of the first ganzfeld-psi experimentation in 1974 there have been over 108 ganzfeld studies encompassing some 2,549 sessions reported in at least forty publications by researchers around the world. Have they all demonstrated psi? Certainly not — hardly any experiments involving human psychology are successful every time. Nevertheless the ganzfeld procedure has seen sufficient successful independent replication that it has come to be viewed as one of parapsychology's best techniques for examining psi under controlled laboratory conditions. Those who have examined the ganzfeld database consider it, as a whole, to be some of the best evidence for a replicable psi effect in parapsychology to date (Utts, 1996). (Rhine Research Center, 2001, page 12)

The major strategy of parapsychology is to declare that by adhering to the strictest canons of scientific methodology, it is possible to (a) demonstrate the existence or nonexistence of psi, and (b) gain knowledge concerning psi (if it exists). Because of this adherence to scientific methodology, parapsychologists claim that mainstream scientific recognition and legitimacy are rightfully theirs. ... an overall review of parapsychological literature (which numbers in the thousands of articles) indicates that psi does indeed exist and even reveals various patterns that seem to be related to its occurrence (Palmer, 1978; Wolman, 1977).

Numerous experiments could be cited as evidence that the existence of psi has been proven. For the purpose of illustration, Beloff (1980) has listed seven experiments that can be considered as highly evidential in support of the existence of psi. (McClenon, 1984, pages 82-83)

I leave the references above for the freshman student who wants to pursue this subject further. It was interesting to me to note that the seven experiments listed by Beloff did not all take place at The Rhine Research Center (formerly FRNM), but were performed at 6 different laboratories in 5 different countries — to answer any insinuations that parapsychological experiments have not been replicated. It should also be noted that, while I have tried to make a case for the evidence of the mental senses, parapsychological research has been conducted by the strictest rules of established physical science, using the evidence of the physical senses only.

As of 1998, Dartmouth College Library had every issue of the *Journal of Parapsychology* going back to Volume #1 in 1937. Anybody could have read it and refuted the claims of this textbook, except that everybody knew, once the social signals had been sent, that they weren't supposed to read it.

CHAPTER 17

Parapsychology and Pseudoscience

In my Harvard Forty-fifth Reunion Report in 2001, I explained that I hadn't given any money to Harvard since 1966, because Harvard supported "scientism," as I called it at the time.

Then it occurred to me that I might give somebody the money that I wasn't giving to Harvard, and the name J.B. Rhine popped into my head. So I looked up J.B. Rhine on the Internet and found The Rhine Research Center. I sent a modest donation, along with a copy of my Reunion Report, to The Rhine Research Center, and as soon as the mail was able to get there, I received an enthusiastic phone call from Sally Rhine Feather, Ph.D., daughter of J.B. Rhine and Executive Director of The Rhine Research Center. They appreciated my donation, and they all enjoyed my Reunion Report, because they all had to deal with the prejudices of scientism every day.

I sent to Dr. Feather, whom I know as "Sally," an accounting of my precognitive dreams, to include in the collection of people's psychic experiences that was started by her mother, Dr. Louisa E. Rhine. And we began an email correspondence.

In one of my precognitive dreams, I had dreamed that the name "Gerald Ford Carter" had historical significance (before there was any hint that either of these men would become President). My usual way of recording dreams is to write the date, followed by "DREAM," and then all the details of the dream that I can remember. But in this case, I printed only the name "GER-

ALD FORD CARTER" in block letters on a piece of paper, with no date and no personal details.

It was an important dream for me personally, with the "Gerald Ford Carter" representing the historical period when those two men were going to be President. I hadn't thought about the scientific value of what I had written until 1999, when I was telling a psychology professor that I had precognitive dreams, and he asked me, "Can you prove that?" Then I realized that I had this name printed in #2 pencil on cheap notebook paper, and if there was some dating method that could prove when it was written, then yes, I could prove it.

I mentioned to Sally that I had this piece of paper which would be evidence of precognition, if there was some dating method that could establish when it was written, and her reply was a real eye-opener for me:

> No, we do not seek to verify or get proof of the veracity or the paranormal quality of any of our reports, as was once the custom back in the old psychical research days. The early psychical researchers went to enormous effort to obtain highly documented cases, and books are filled with excellent cases that could stand for as much scientific proof as case material could ever be. And you know what? No one in the scientific world seemed to pay much attention. (Not that they pay so much attention to scientific research results [in parapsychology] either, I hasten to add). (from an email of September 20, 2001)

So this was the kind of complete and total prejudice that parapsychology had to put up with. Refusal to look at the evidence was about as unscientific as a scientific establishment could be, except for depriving people of their livelihood. And I was going to find out that they were doing that, too.

In 2009, at age 75, I moved to Durham, North Carolina, to work as a volunteer at The Rhine Research Center. I figured I could be useful in my old age doing clerical work, and I became Business Manager of the *Journal of Parapsychology*, mostly keeping track of subscribers.

I have also received an education in parapsychology, reading a great many books on the subject, taking a course in it, and attending most of the Friday night presentations at The Rhine, a few conferences, meetings of the Psychic Experiences Group

(PEG) and the Rhine Book Club, and a couple of spoon-bending parties. I have been witnessing here the cutting edge of our culture, whether or not the scientific establishment and the universities are aware of it.

Some of the most impressive presentations were William Bengston showing how he cured 200 mice of cancer, Daryl Bem demonstrating how things learned in the future can influence our knowledge in the present, Ginette Nachman on synchronicities, those incredible coincidences which defy all laws of probability, and a report from a hospital in India where they pray for their patients, with statistically significant results. But my subject here is not the wonderful things that are being discovered in parapsychology, but dirty science. That's why I was attracted to The Rhine Research Center in the first place, to oppose the bias and prejudice directed at parapsychology, and that's what I found myself doing when I got here.

In one of my first encounters at The Rhine, a man came in with his son, whose middle-school science project had received a failing grade because the subject he had chosen was parapsychology. This seems cruel, but I suppose it is all part of the process of learning the politics of physicalism, because a few years later I met a person with a PhD degree who had been fired from a major university because of too much interest in parapsychology. Through my various encounters with people at The Rhine, I learned that dirty science was more than just ridicule, but had the power to deny people publication, funding, and employment.

And then They turn around and say that the reason for a declining number of papers on parapsychology in mainstream science journals is because of a declining interest because the subject is bogus.

Knowing my interest in dirty science, Sally asked me if I could try to persuade Wikipedia not to label parapsychology a "pseudoscience." She said that Nancy Zingrone, Ph.D., former President of the Parapsychological Association, had tried that with no success. If the person with the highest status in the field of parapsychology had failed, there didn't seem to be much hope for me, but this was what I had come here to do, and I was willing to try.

According to Wikipedia, parapsychology had been labeled a "pseudoscience" because no evidence had been found to substantiate the existence of psychic phenomena in more than a century of research. That claim, of course, was absurd. I had access to literally a ton of evidence to refute it, in the Alex Tanous Library at The Rhine Research Center. Where did anybody get the idea that there was no evidence? I quickly discovered the reference to the National Academy of Sciences. I want to quote it again, and make a spectacle out of it, because it is so precious. While there is some debate over whether anybody actually asserted that nothing heavier than air could fly, or that the Patent Office should be closed because nothing new could possibly be invented, this assertion is fully documented, and, as I have already said and I am happy to repeat, will be recorded in the annals of folly forever:

> The committee finds no scientific justification from research conducted over a period of 130 years for the existence of parapsychological phenomena. (Druckman and Swets, 1988, page 22).

I found the report on Abebooks.com, saw that only one chapter and five references were devoted to parapsychology, and saw the bias with which it was written, picking on the weakest studies and representing them in the least favorable light, with smear-tactics such as allegorical allegations of "dirty test tubes" and ridicule for things like spoon-bending. It was not at all scientific. It was one of those publications that should be assigned to college freshman classes in critical thinking, as an example of illegitimate arguments.

My strategy, then, since Wikipedia appeared to be intractable, was to approach the National Academy of Sciences and let them know that this highly biased, highly inaccurate, and actually sleazy report had been issued in their name. At Harvard we had learned, as part of the in-group attitude, not to be impressed by anybody, but as I read the biographical data of the Councilors, I was deeply impressed and truly in awe of the stature of these people. There was a huge contrast between the highest quality of science that these people represented and the inaccurate and irresponsible report that now carried their name and their reputation with it.

So it was with the highest possible respect that I decided to write a letter to the President of the National Academy of Sciences, to make him aware of the travesty that had been issued in the name of his organization. I turned for help in composing this letter to Sally Rhine Feather, Ph.D., Executive Director of The Rhine Research Center, and John Palmer, Ph.D., Editor of the *Journal of Parapsychology*. Both of them were very skeptical that my letter would accomplish anything, based upon years of experience in dealing with the prejudice.

But they both helped me enormously in composing the letter. Sally advised me not to use words like "sham" and "farce," and actually supplied some of the wording for the letter. Her style was much more genteel than mine, and more appropriate for approaching a person with as high a status as the President of the National Academy of Sciences.

John Palmer informed me that he, along with Charles Honorton and Jessica Utts, three people expert in the field of parapsychology, had already published a rebuttal to the report of the National Academy of Sciences (Palmer et al., 1988). He let me make a copy of his copy, because the publication was not easily available. One of the many things the rebuttal pointed out was that the two principal evaluators of parapsychology, Ray Hyman and James Alcock, were known zealots against parapsychology. And there was nobody on the committee expert in parapsychology, to counterbalance their efforts and perhaps educate the committee on the subject.

Jessica Utts, Ph.D., Chair of the Statistics Department at the University of California Irvine, backed up my letter with a letter of her own, on a more professional and expert level.

I want to reprint here my letter to Ralph J. Cicerone, Ph.D., President of the National Academy of Sciences, because it is the best statement I could possibly make, and it represents itself exactly:

March 14, 2011

Ralph J. Cicerone, President
National Academy of Sciences
500 Fifth Street, NW
Washington, DC 20001

Dear Dr. Cicerone,

I have received no reply to my letter of January 21, 2011, from anybody in your organization, so I am sending the whole letter again.

When I look up "parapsychology" on Wikipedia, I find that parapsychology has been labeled a "pseudoscience," as follows:

> External scientists and skeptics have criticized the discipline as being a pseudoscience because, as they see it, parapsychologists continue investigation despite not having demonstrated conclusive evidence of psychic abilities in more than a century of research.[20][21][22]

And when I look for the source of that claim, I find the following statement:

> In 1988, the U.S. National Academy of Sciences published a report on the subject that concluded that "The committee finds no scientific justification from research conducted over a period of 130 years for the existence of parapsychological phenomena." (1)

When I actually read the report, I found that the committee DID NOT LOOK AT the research of 130 years. They cite only 5 references to the *Journal of Parapsychology*, all from the 1980s, and only two on actual research, one reference from the *Journal of the American Society for Psychical Research*, and no references at all from the peer-reviewed British and European journals. They make no mention of the name "J.B. Rhine," let alone refute his research or refute the fact that his findings have been replicated by a great many independent laboratories in many countries throughout the world. (2,3,4,5) And the figure of "130 years" seems to have been plucked out of thin air.

Any statement I make at this point would be an understatement: The scope of the study was insufficient to support its sweeping conclusion. In order to demonstrate with a 95% probability that not a single experiment yielded evidence of parapsychological phenomena, the committee would have had to review 95% of the experiments in the field and refute each of them accurately. Obviously the scope of this study fell far short of that.

This was the first of the criticisms at the time in a rebuttal paper by three of the most highly qualified people in the field of parapsychology. (6) They made many more valid criticisms, which in my opinion seemed to turn the issue into a debate. Some of these points may be debatable, but the mathematics of science is generally accepted: A certain sample-size is necessary to claim certain results, and this study had only a tiny fraction of the sample-size necessary to support its sweeping claims.

I believe that most people at the time accepted the conclusion based on the high status and reputation of the National Academy of Sciences, and never bothered to read the report. Some writers have had to create the necessary fiction, saying that the committee did do the complete study, just to make the finding plausible to a general audience. Robert L. Park, in *Voodoo Science*, said:

> In 1987, at the request of the U.S. Army, the National Academy of Sciences undertook a complete review [sic] of all the literature on parapsychology as part of a larger study of unconventional methods of enhancing human performance. (7)

And, from http://www.skepticfiles.org/mys1/armypsy.htm —

> In arriving at that conclusion, the panel reviewed research literature dating back 130 years [sic] ...

Obviously, this report, using the prestigious name of the National Academy of Sciences, has done enormous damage to the field of parapsychology in the United States. (Parapsychology is alive and well in the United Kingdom.) What may be not so obvious is the damage that this report can potentially do to the reputation of the National Academy of Sciences, that such a blatantly invalid report can pass through the scrutiny of the scientifically trained leadership of the National Academy of Sciences. Since this didn't happen on your watch, you aren't responsible for it, but it would be to your great credit if you would look into this matter with a view to correcting the serious omissions and premature conclusions.

I am trying to approach you here in a friendly way, hoping that you will recognize that it would be to the benefit of both the National Academy of Sciences and parapsychology to reverse this judgment. I am not asking you to go out on a limb and endorse parapsychology. I am only asking you, as a scientist, to recognize that the committee's sweeping conclusion was invalid, and to issue a public statement to that effect.

Sincerely,

Robert S. Gebelein

Copies to:

Sally Rhine Feather, Director, The Rhine Research Center
John Palmer, Editor, Journal of Parapsychology
Parapsychological Association, Inc.

References:

(1) Druckman, D. and Swets, J. A. eds. (1988). *Enhancing Human Performance: Issues, Theories and Techniques*. National Academy Press, Washington, D.C.. p. 22.

(2) Utts, J. (1996). An Assessment of the Evidence for Psychic Functioning. *Journal of Scientific Exploration*, Vol. 10, 1, p. 3-30.

(3) Palmer, John. "Extrasensory Perception: Research Findings," in *Advances in Parapsychological Research, 2: Extrasensory Perception*, Stanley Krippner, ed. New York: Plenum, 1978.

(4) Wolman, B.B., ed. *Handbook of Parapsychology*. New York: Van Nostrand Reinhold, 1977.

(5) Beloff, John, "Seven Evidential Experiments," *Zetetic Scholar*, 6 (1980), 91-94.

(6) Palmer, J., Honorton, C., and Utts, J. (1988). *Reply to the National Research Council Study on Parapsychology*. Parapsychological Association, Inc.

(7) Park, Robert L. (2000). *Voodoo Science: The road from foolishness to fraud*. Oxford University Press. p. 197.

I was delighted even to receive a reply from Dr. Cicerone. It was a very good letter, worthy of a person of his stature. In it he offered to find a "knowledgeable disinterested" person to investigate the subject. I couldn't help thinking that that might be as difficult as finding a disinterested person on the subject of psychedelic drugs or the Presidential election of 2000.

I didn't hear any more from Dr. Cicerone or the National Academy of Sciences, but a couple of years later I looked up "parapsychology" on Wikipedia, and it seemed that the reference to the National Academy of Sciences had been deleted. When I looked at it again ("last modified on 15 January 2016"), I could see that the quote had been moved down from its prominent place to page 15, and the reference number had been moved down from 20-22 to 149.

I speculated that the National Academy of Sciences might have applied some pressure on Wikipedia to drop the reference, but Dr. Cicerone informed me that this didn't happen. So by some miracle Wikipedia chose to move this reference, really their main support for their definition of "pseudoscience," to a less conspicuous place.

As of January 15, 2016, the definition of "parapsychology" in Wikipedia began with the words, "Parapsychology is a pseudoscience," whereas previously that assertion wasn't made until the second or third sentence. And there was still the statement in the second paragraph, "Parapsychology has been criticised for continuing investigation despite being unable to provide convincing evidence for the existence of any psychic phenomena after more than a century of research." But the references had changed to Hyman, Flew, Cordón, Bunge, Blitz, and Stein, persons whose status was not comparable to the National Academy of Sciences. I was wondering why they would give up such an authoritative reference, but I was thankful for it.

So some small progress had been made. I learned, when I was dealing with boulders on my New Hampshire property, that if I could move it a quarter of an inch, I could move it across the yard. I hope that the same principle will apply here.

Didn't the editors of Wikipedia know that parapsychology had been recognized as a science by the American Association for the Advancement of Science (AAAS) since 1969? Yes, they did, but they presented it in an unfavorable way, with no acknowledgement that the members of that organization, all scientists, voted by a 5 to 1 majority to recognize parapsychology as a science, but emphasizing instead the one zealot who called it a "pseudoscience:"

> In 1969, under the direction of anthropologist Margaret Mead, the Parapsychological Association became affiliated with the American Association for the Advancement of Science (AAAS), the largest general scientific society in the world.[48] In 1979, physicist John A. Wheeler said that parapsychology is pseudoscientific, and that the affiliation of the PA to the AAAS needed to be reconsidered.[49][50] His challenge to parapsychology's AAAS affiliation was unsuccessful.[50] (Wikipedia, 15 January 2016, page 6, references 48-50)

This has been the general tone of Wikipedia on parapsychology — to minimize, discredit, slant, smear, insinuate, and accuse of incompetence, cheating, or fraud — in one word, to bias. A large proportion of their references has come from people known to have a bias, such as James Randi, Ray Hyman, James Alcock,

Paul Kurtz, *Skeptical Inquirer*, CSICOP, or Prometheus Books, or in titles such as *Voodoo Science* that have advertised a bias.

Over and over again these people and other detractors of parapsychology have described themselves as "skeptics," to convey the impression that they have no set opinion and are open to scientific evidence, when actually they are "pseudoskeptics," meaning that they do have a set opinion and are closed to scientific evidence to the contrary, and will find any way they can to explain away the evidence.

I know that psychic and spiritual phenomena exist because I have experienced them myself. I don't need any scientific proof. And then there are people who know for certain that psychic and spiritual phenomena do NOT exist because they have NOT experienced these things themselves. And they will NOT accept any scientific proof. What's wrong with that?

I have learned from several sources that Wikipedia is inaccurate. From John Kruth, the present Executive Director of The Rhine Research Center, I learned that there are people called "Guerrilla Skeptics" who are actively making it inaccurate by imposing their ideological bias upon it. I tried to look up "Guerrilla skeptics" and "Guerrilla skepticism" on Wikipedia and found that these pages did not exist. I had to look up "Susan Gerbic," the founder and organizer, to find the section, "Guerrilla Skepticism on Wikipedia (GSoW)."

Susan Gerbic, it said, "grew up ... as a Southern Baptist," and she said, "I never heard the word atheist until I was in my late teens." It didn't say that she swung from the one extreme to the other. This has to be inferred when she says she found other so-called "skeptics" and was educated by the James Randi Educational Foundation.

Yes, James Randi has actually established an Educational Foundation, where he has offered instruction in critical thinking. It had 68,853 likes on Facebook when I looked it up. Judging from his book, *Flim-Flam!*, I can imagine what kind of "critical thinking" is being offered. And I can imagine what Susan Gerbic has been doing on Wikipedia "to improve content."

[I have to talk about Wikipedia in the past tense because the content keeps changing. As of August 29, 2018, they no longer started off with "Parapsychology is a pseudoscience," but the general derogatory tone continued. They still seemed to favor people with a known bias, and their references to the National Academy of Sciences were numbers 96 and 159.]

Some extremists, in a supreme effort to discredit parapsychology, even changed the math, to deny the reality of Daryl Bem's findings in his paper, "Feeling the Future" (Bem, 2011).

In the fall of 2010, when Daryl Bem came to The Rhine, he did a little experiment with his audience. First he flashed on the screen a series of words, 24 in all. Then he asked us to write down all we could remember.

I didn't remember very many, only 9 out of 24. As a child I did much better at memory games, but I already knew that my short-term memory was failing in my old age.

After we had written down all the words we could remember, Bem gave us instruction on half of them. He divided them up into 4 categories — clothes, fruits, occupations, and animals — with 3 words in each category. He then asked us to check off these words on our lists, and count the number of words we had checked off, and compare that to the number of words that we hadn't checked off, that we hadn't had instruction on.

To my amazement, I had remembered 7 of the words we had received future instruction on, compared to only 2 that we hadn't had instruction on. He asked how many of us remembered more of the words that we had had future instruction on, and most of the people in the audience raised their hands. In other words, we had the test first and the instruction afterwards, just the reverse order of the way it usually happens.

That's basically what his paper "Feeling the Future" was about. There were two experiments with future instruction like the one we had experienced, but with 48 words ("Retroactive facilitation of recall"). There were also two experiments on "Precognitive approach/avoidance," the first "Detection of Erotic Stimuli" and the second "Avoidance of Negative Stimuli," two experiments on "Retroactive priming," two experiments on "Retroactive habituation," and one experiment on "Retroactive Induction of Boredom," for a total of 9 experiments in all. He claimed that 8 out of the 9 experiments were statistically significant, in favor of being able to perceive the future. When you combine them all, the result is impressive. He calculated the probability $p = 1.34 \times 10^{-11}$ (about one in 75 billion) that this all happened by chance.

All these experiments were run by computer. The experimenter left the room and the participants interacted solely with the computer. The software is available both for Windows and Macintosh, so that these experiments can be precisely replicated. When Bem returned to The Rhine a couple of years later, he said that the experiments had been replicated by 80 independent observers in 30 countries.

This is all a gross oversimplification of a paper which, in my printout, is 61 pages long, double-spaced. In it Bem explains how he ruled out the possibility of clairvoyance, if the random number table is set, or psychokinesis, if random numbers are generated dynamically, by running experiments both ways with no bias shown either way. On the erotic pictures, he allows people four sexual preferences. In some experiments he separates people into extraverts and introverts, and in the second habituation experiment he separates out "erotic stimulus seekers," to see if these subgroups would differ in their psi performances. They do. He has taken many factors into consideration in what seems to me to be a very complicated set of experiments.

One factor he mentions at the end is the experimenter effect. Even though all the interaction during the experiment is with the computer, any indication beforehand that the experimenter wants negative results can bias the experiment and make it come out negatively or with "chance" results. On the other hand, something the pseudoskeptics don't seem to understand and needs to be pointed out is that even the most positive bias exerted by an experimenter who is a "believer" is not going to give the participant the ability to perceive the future if that ability does not exist. Experimenter psi, of course, can create positive results, but then that would demonstrate that psychic abilities are real.

Some of the criticism against this paper was predictable. James Alcock, a known zealot against parapsychology, published a critical piece in *Skeptical Inquirer*, a magazine with a known bias against anything psychic or spiritual (Alcock, 2011). As I have said before, I don't see *Skeptical Inquirer* (or CSICOP, now "CSI," or Prometheus Books) as either scholarly or scientific. They are dedicated to supporting a belief system — physicalism. In my opinion, they are psychologically interlocked in a struggle with the religious right, on that level.

As if the bias represented by Alcock in *Skeptical Inquirer* wasn't known immediately, he advertises it right at the beginning by saying, "Parapsychology has long struggled, unsuccessfully, for acceptance in the halls of science."

I think I have to inform James Alcock, because the rest of us know, that parapsychology was accepted in the halls of science in 1969, when the Parapsychological Association (PA) was accepted into the American Association for the Advancement of Science (AAAS) by a 5-to-1 vote. I'm sorry I have to keep repeating this over and over, but people keep denying it over and over. The majority of scientists accepted the reality of psychic phenomena in 1994, according to a survey by Bem and Honorton. Only a minority of individuals such as Alcock refuse to accept it.

Alcock then goes on to dismiss all of J.B. Rhine's work because of "methodological problems," and Helmut Schmidt's work because of "methodological errors" and "methodological flaws." The problem here is that he doesn't identify what any of these errors are. This is a common tactic of people in the putdown business. They get to implant the suggestion in people's minds that there are errors, and with it the insinuation that these errors invalidate the entire work, without actually specifying any errors. If Alcock is trying to do science here, he is the one with the "methodological errors."

Alcock's main complaint about Bem's experiments is that they are confusing. Yes, they are complicated, but in my opinion Alcock is the one creating the confusion. He uses words like "messy" and "muddle" to communicate his bias, which we all know already.

He makes a factual error in describing the first experiment, in which Bem switches from 2 groups of 18 pictures to 3 groups of 12 pictures. He doesn't seem to register that the problem for the participant is always to guess which of 2 curtains the picture is behind, and not which group the picture belongs to.

He jumps on Bem's two-question tests of extraversion/introversion and "erotic stimulus seekers," but actually these separations have no bearing on the main finding of the study, only on which kinds of people do better at "feeling the future."

And so on — Alcock goes on and on, jumping on every irregularity with appropriate smear tactics to make Bem look bad. But we know Alcock's bias, and we know that this is what he is going to do. And it gets boring.

Not so boring is the paper by Eric-Jan Wagenmakers, Ruud Wetzels, Denny Borsboom, and Han van der Maas, "Why Psychologists Must Change the Way They Analyze Their Data: The Case of Psi" (Wagenmakers et al., 2011). They are so certain that precognition does not exist that they are saying we have to change the math so that experiments with psi produce results consistent with their belief.

The math that they want to change is almost as established as the scientific establishment itself. The 5% probability level for statistical significance has become part of the rules of science since first proposed by Ronald Fisher in 1925. This is an arbitrary cut-off point, but if there is any doubt that the one chance in 20 might really have been arrived at by chance, you can repeat the same experiment, and if you get the same results again, then the probability is one in 400 that they both happened by chance. Or you can run 9 experiments, as Daryl Bem did, and arrive at a probability of 1.34 in 100,000,000,000 that it all happened by chance.

The binomial distribution that determines these probabilities goes back even farther than that, to Swiss mathematician Jakob Bernoulli, who died in 1705. Mathematics is the nearest thing to absolute truth that we have. Changing the math would be like changing the rules of science itself, just because one scientific finding didn't agree with somebody's belief system. And if the math was changed, it would affect, retroactively, every scientific finding that was ever determined using the old math.

It is a testimony to the quality and the paradigm-changing nature of Bem's experiment that these people want to change the math in order to refute it.

Wagenmakers et al. interpret Bem's results using Bayesian statistics instead of the established statistical methods. In Bayesian statistics, you have to assign probability figures to the false positive and the true positive. They are so certain of their belief that they assign a probability of 10^{-20} to the possibility that psi phenomena are real. This is an amazing degree of certainty. My usual illustration of certainty is the probability that the sun will rise tomorrow. But their figure of 10^{-20} is the equivalent of the certainty that the sun will rise every day for about 274,000,000,000,000,000 years, or about 20,000,000 times the age of the universe.

Then, given this near-zero probability figure for the true positive, Wagenmakers et al. are able to calculate a near-zero result (19×10^{-20}) for the reality of psi. In other words, in Bayesian analysis, they are able to plug their bias into the equation, and out the other end comes an answer that reflects their bias. What's wrong with that?

The old-fashioned statistics is simply a mathematical calculation, as I said, the closest thing to absolute truth that we have. It simply tells us where the result lies on a probability curve, without somebody's opinion being able to bias the answer. Therefore I would call it more scientific.

While the arrogance of Wagenmakers et al. is novel and entertaining, on the serious side it illustrates the extreme bias that is operating here: Any physical explanation is preferable to any explanation that doesn't conform to known physical laws. To enforce that rule, extreme ridicule is applied. If the extreme ridicule doesn't force people to conform, then they are deprived of their entire professional and social standing. None of this has anything to do with science. It is strictly social manipulation and domination. Wagenmakers et al. have added a new wrinkle by trying to support this extreme bias with mathematics.

They don't do as much of the boring stuff as James Alcock. Their argument at least gives the appearance of being scientific, although they do mention that "nobody has ever collected the $1,000,000." Their method seems to be the quick fiction, grasping at straws, looking for any legitimate argument that might discredit Bem's findings.

They seem particularly concerned that if people were psychic, the casinos might go bankrupt. Actually, I have heard that the casinos take care of themselves, identifying the people who win too much too often and inviting them not to come back. But if you look at the people who are very rich or very "lucky" or more successful than you are for no apparent reason, how many of them have psychic abilities? They aren't saying.

Daryl Bem has said, "Extraordinary claims require extraordinary evidence." This is one point upon which he and Wagenmakers et al. seem to agree. But I agree with John Palmer (1987) that this exerts a bias. The same rules of science should apply to all claims, whether or not somebody thinks they are extraordinary, because if you think a claim is extraordinary, that implies that you think you know something about the phenomenon in question that makes it extraordinary, and that would be a bias.

Wagenmakers et al. do make a valid point, though, when they say that the evidence would have to be much stronger to convince THEM of the reality of psi phenomena. But this is a psychological or sociological problem, of overcoming the bias of a major subculture, and not simply a problem of demonstrating the reality of psi phenomena. The reality of psi phenomena has been demonstrated over and over again for the impartial observer, by J.B. Rhine and many others.

If 90% of Americans have had a psychic experience, then why is this such an extraordinary claim? I didn't hear much about psychic abilities until I became aware that I was having precognitive dreams myself and began telling people about it. Then people gushed forth with their own psychic experiences. It seemed that everybody had a story to tell. Why were they reluctant to tell their stories unless they knew they were talking to a true believer?

The need for secrecy goes back a long way. The Catholic Church and Protestant churches alike burned people at the stake for witchcraft. Having psychic abilities was considered the work of the Devil, or witchcraft. Even today, religious people shy away from psychic abilities because they believe these things are the work of the Devil. So there has been a long history of religious persecution, causing people to keep these things secret, even before the scientific era.

And then when science became dominant, people who had psychic abilities were lumped in with people who heard voices or had hallucinations — they were considered to have a mental disorder. So of course people have been secretive about it.

These claims are only "extraordinary" because they have been suppressed for so long by authorities who have the power to put you to death, condemn you as possessed by the Devil, or commit you to an insane asylum. And now psychic experiences are considered extraordinary only because the scientific establishment keeps rejecting the ordinary evidence presented, thousands of times, demanding extraordinary evidence. This rejection of evidence is, in itself, evidence of something. There is a force beyond chance operating here. That force is the extreme bias, which actually destroys science and makes scientists lose all credibility in this area.

That bias is taught to the two million students who graduate from college every year in America. I was interested to read in *Mindfield: The Bulletin of the Parapsychological Association* a paper by Chris A. Roe entitled "What Are Psychology Students Told About the Current State of Parapsychology?" It confirms what I observed in the one college textbook and extends it to other college textbooks. Of the 8 introductory psychology textbooks that he found in his university (UK) library, 4 of them said nothing about parapsychology, and the other four had the same kind of disparaging unscientific remarks and insinuations about it that I had observed, using words like "fraudulent," "threats to the validity of research," "vulnerable people are being taken advantage of," "more troubling," and "questionable use of statistical analysis," and cite as their authorities known sources of bias *Skeptical Inquirer*, CSICOP, James Randi, and James Alcock.

Roe also refers to a study done from 1990 to 2002 by James McClenon, Miguel Roig, Matthew D. Smith, and Gillian Ferrier (McClenon et al., 2003), analyzing 64 textbooks published in the 1980s, 52 textbooks published in the 1990s, and 57 textbooks published for fall 2002. These authors observed that "The 2002 texts tended to portray a more skeptical tone." Their analysis of the 2002 texts revealed:

> ... Authors reviewing experimental evidence supporting of belief in psi (Bem & Honorton, Honorton, McConnell, Rhine) were cited collectively 39 times. Authors skeptical of these claims (Milton & Wiseman, Hyman, Randi, Blackmore, Hansel, Alcock, Marks, Swets & Bjork) were cited collectively 94 times. Skeptical positions received roughly 2 times more coverage than parapsychological claims.

... Authors publishing in the popular, nonrefereed *Skeptical Inquirer* were cited 58 times, whereas those publishing in the *Journal of Parapsychology* were cited only 22 times. (McClenon et al., 2003, pages 172-3)

This is in textbooks. I can't think of anything else I have in a psychology textbook, except for attempts to dismiss Freud and Jung, that isn't based 100% on scholarly references.

Also McClenon et al. use words that suggest that litigation might be possible against some of these textbook publishers. They use the word "misrepresentation" (p. 168), also "Alcock's (1990) false and misleading statements" (p. 171), and also critics' accusations of "fraud" (p. 175) by experimenters. I am thinking that lawyers might take an interest in this. Also James Randi should not be the judge of whether his million-dollar offer should be paid. It should be taken to an impartial court of law to decide.

McClenon et al. conclude by saying:

... Science is a political and rhetorical process. Although parapsychologists may feel that the evidence they generate supports belief in psi, political and rhetorical factors affect the treatment that this evidence receives. (McClenon et al., 2003, page 176)

Yes, I agree that political and rhetorical factors influence the way the evidence is treated, but I don't agree that politics and rhetoric are a part of science. Science uses accurate observation, along with accurate logic and accurate mathematics, to explain those observations. The politics and rhetoric are part of the corruption of science, and if they are allowed to continue to dominate, they will reduce the accuracy of whatever calls itself "science" back down to the level of knowledge of the Stone-Age chieftain who rules everybody with a club.

I think the concluding statement of Nancy Zingrone, in her Presidential Address to the Parapsychological Association in 2001, is more accurate and more appropriate:

... If we add an understanding of the social, political, and rhetorical surround to the methodological and analytical tools we already have on our research bench, progress in parapsychology is inevitable. (Zingrone, 2002, page 25)

"The social, political, and rhetorical surround" is not science. If it poses as science, it is what I call "dirty science." It is pseudo-science. For the sake of the two million Americans who graduate from our colleges and universities each year, let's get this pseudo-science out of the workplace.

CHAPTER 18

Cleaning up Science

Physicalism, the assertion that there is no reality beyond the physical or what can be explained by known physical laws, dominates the academic community as if it were a hypnotic command. It is only an assertion. It has not been proved scientifically, and logically cannot be proved without a complete knowledge of everything that exists. It has actually been disproved by the billions of people worldwide who have had psychic and spiritual experiences. And yet it is accepted at American colleges and universities as a first principle, an axiom of rational thought that is considered so absolutely true that those who even question it are dismissed as mentally incompetent.

Evidence of the domination of physicalism is the fact that precognition, telepathy, clairvoyance, remote viewing, psychokinesis, energy medicine, spirit entities, the power of prayer, reincarnation, levitation, or intelligent design have not been studied or taught at ranked American colleges and universities except in low-profile situations or in special cases which all have explanations as to why they were allowed.

The domination of physicalism is blocking our cultural advancement in the areas of the mental, the psychic, and the spiritual. Our knowledge of these non-physical aspects of human existence must be developed if our species is to survive at all. Our physical and technical knowledge has raced ahead of our understanding and development of the human being. Einstein's discovery of the force that threatens to destroy us was developed into "overkill" in about 50 years, whereas Freud's discovery of the self-knowledge that might save us, a discovery that was made before Einstein's, has bogged down in political wrangling, largely due to physicalism.

Instead of being supported by any scientific or legitimate arguments, physicalism is defended and enforced by unscientific and illegitimate arguments, such as ridicule, authoritarian pronouncements, refusal to look at the evidence, intimidation, and power politics. Freshman students of critical thinking learn the difference between legitimate and illegitimate arguments. Therefore academic people with PhDs also know the difference. But they are controlled by the hard ridicule: People who express an interest in non-physical subjects are deprived of publication, funding, and employment. Tenured professors who can't be fired are shunned.

We, the people, can recognize these unscientific arguments for what they are, and not let ourselves be manipulated by them. But within the academic community, the hard ridicule seems to be backed by a consensus of the whole social group, across every ranked college and university in the United States. It does no good for an individual to recognize the unscientific arguments if the individual violates this consensus in so doing. The consensus is more powerful than any individual or any rational argument.

As a result, while qualified and responsible people are studying the psychic and the spiritual and building up a body of knowledge outside of the academic community, our major educational institutions, having become paralyzed by the domination of physicalism, are falling farther and farther behind the actual cultural knowledge.

So the question is, who needs the academic community? Can't we just proceed as we are going, exploring and studying and expanding our horizons independently of the control of the academic community?

We might do that, but as long as physicalism dominates the academic community, it dominates the whole culture. The academic community preserves and teaches the established culture. The academic community defines that culture. The academic community decides what is "established." The academic community decides what the culture recognizes as "knowledge." The academic community is accredited. The academic community issues the credentials. The academic community in the United States confers degrees every year on roughly two million people, who then go out into the world and propagate the culture according to what they have been taught. If the academic community is dominated by dirty science, so is the culture.

Yes, there are a great many individuals and groups studying and teaching shamanism, the siddhis, and all the other psychic and spiritual subjects that I have named. And I personally claim to have designed a new civilization. But unless these subjects are accepted by the academic community, they will not be preserved as part of the cultural knowledge. They will just continue to be part of the "occult," the "hidden." The academic community needs to change, to recognize the legitimacy of all these subjects which have been ignored, dismissed, and/or suppressed because of the domination of physicalism.

With that thought in mind, I sent my letter in 2013 to the presidents and chancellors of the highest ranked 137 American colleges and universities, asking them, as persons of the highest status within the academic in-group, if they would exert their influence to end the domination of physicalism. In that letter, I made three suggestions which I believe would solve the problem:

1. All subjects should be debatable. No subject should be career-ending.

2. Scientists in any particular field should be considered the best qualified to operate in that particular field, and should be considered the best qualified to determine what the rules of evidence should be for that particular field. For example, physical scientists should not impose their rules of evidence on people studying mental processes.

3. Lawyers who use illegal methods are disbarred. Similarly, scientists who use unscientific methods should lose their credentials. They should at least follow the rules that are taught in freshman courses in critical thinking. Science needs to police itself, to protect those of lower status in the in-group and protect the public, who are less informed. To start with, the scientists themselves should understand that they lose credibility when they use unscientific methods.

In the presidents' polite replies, I didn't get much information. One agreed with me that physicalism was a problem. Two disagreed with me. Five agreed that all subjects should be debatable, but four out of those five gave me the impression that they thought that all subjects were already debatable at their institutions. Four of them informed me that the faculty had total freedom and authority to determine what was studied or taught at their institutions.

There is no indication in any of their replies that they are aware that there is anything unscientific or unscholarly going on, especially such a thing as academic people being denied publication, funding, or employment because of nonconforming interests or activities. Those who told me that individual faculty members were free to choose the curricula missed the point I was trying to make that ordinary faculty members were NOT free, because of the conformity forced upon them by physicalism, and that I was approaching the persons with the highest status in the in-group because I felt that they had more freedom to deviate from that conformity.

Maybe they felt insulted that I called the academic community an "in-group." I had no way of knowing that from their very polite letters. I can't cite any scientific studies showing that the academic community is an in-group, so by my own rules, until such a study is conducted, I suppose I can't expect the academic community to recognize how all of academia functions as one homogeneous in-group, how the in-group thinking is backed by prejudices and illegitimate arguments, and just how powerful an influence this is on the academic community and our cultural knowledge in general. Could there ever be such a study, or would it be too close to home for academic people to study their own colleagues? It seems that some self-knowledge is in order.

If individual faculty members determine what is studied and taught, then a university can control what is studied or taught by employing only faculty members who have their degrees in approved subjects. If the hiring and firing is done at the department level, then the university at some higher level is responsible for determining what departments there are, and which areas of knowledge they cover. At this higher level subjects such as the psychic and the spiritual are excluded, and the individual faculty members, who all have their degrees in approved subjects, never have to give such out-group subjects a thought. As I see it, presidents of universities should know this, and they are just giving me a polite bureaucratic runaround. This selective selection process ensures that a person can't get a degree in parapsychology, much less a PhD, at any ranked college or university in the United States, and thus helps to perpetuate itself.

Perhaps there is no contamination of freedom of inquiry or freedom of debate because the selective hiring and firing (or shunning of tenured professors) keeps unwanted opinions out of the academic environment. The social pressures keep undesirable people out of the in-group in the first place.

Nobody commented on my suggestion that the scientific community needs to police itself. This might seem to be a restriction of academic freedom. But freedom should not include freedom to take away another person's freedom. In the real world we have police to protect people's freedom. Are people in the academic world "above" all that?

The replies to my letter reflected the same innocent view of the academic environment that I had while I was at Harvard — freedom of inquiry, freedom to think for oneself, and freedom of debate. In my college days the thought never entered my mind that academic people might have had prejudices. I was aware of a prejudice against football, but that was somehow OK, probably because football wasn't academic. And I was aware that people with serious religious beliefs were ridiculed unmercifully, but that, too, seemed OK, probably because rigid religious beliefs were a serious violation of "academic freedom" and "freedom of inquiry." Prejudice was something religious people had, not academic people.

But as I grew older, I heard the various prejudices on The List echoed by various academic people. I didn't realize how serious they were until I became involved in the field of parapsychology and learned that people had actually lost their jobs because of their interest in out-group subjects. Being denied publication, funding, and employment and being shunned by one's peers is an extreme case of how the norms of a social group can restrict one's freedom, reminiscent of religious cults.

But these letters from the leaders of the academic community were reflecting the innocent view that I had when I was in college. Am I deluded in believing that there are sinister social forces at work to limit and corrupt academic inquiry?

I don't think so. There is evidence to support my view — literally a "ton" of evidence at The Rhine Research Center. The commission appointed by the National Academy of Sciences completely denied all this evidence when they concluded in 1988 that nothing had been discovered in the field of parapsychology. Similarly, Wagenmakers et al. had to deny a large chunk of this evidence in order to claim a near-zero probability for the existence of precognition. I would say that there is a near-zero probability that precognition does NOT exist, based on my "GERALD FORD CARTER" dream and a dozen others.

The only way it can rationally be argued that parapsychological phenomena do not exist is to start off with the assertion that these things do not exist. The same is true for ghosts and angels. From that primary assertion, then, all evidence of such phenomena is assumed to be faulty. Science has not demonstrated that these things do not exist. It has simply ignored them, as established science has focused on the physical. That gap is obvious and needs to be recognized.

The view of a purely physical universe is supported by illegitimate arguments that can easily be recognized by freshman students of critical thinking and therefore should be recognized by people with PhD degrees. And yet the illegitimate arguments of people such as James Randi are actually cited in college textbooks in defense of physicalism.

No, I am not deluded when I see prejudice and corruption infecting the academic community. And I don't think I am deluded when I see the leaders of that community still showing the same innocence that I had when I was in college. Or maybe they just aren't admitting that I am right.

Bertrand Russell set the tone for the 20th century with this 1903 statement expressing his near-certain belief that there is no reality in the universe beyond the physical:

> That man is the product of causes which had no prevision of the end they were achieving; that his origin, his growth, his hopes and fears, his loves and his beliefs, are but the outcome of accidental collocations of atoms; that no fire, no heroism, no intensity of thought or feeling, can preserve an individual life beyond the grave; that all the labours of the ages, all the devotion, all the inspiration, all the noonday brightness of human genius, are destined to extinction in the vast death of the solar system; and that the

whole temple of Man's achievement must inevitably be buried beneath the debris of a universe in ruins—all these things, if not quite beyond dispute, are yet so nearly certain, that no philosophy that rejects them can hope to stand. Only within the scaffolding of these truths, only on the firm foundation of unyielding despair, can the soul's habitation henceforth be safely built. (Russell, 1923, pages 6-7)

Russell's near certainty has become a certainty, accepted by academic people as a first principle, like a religious belief, as reflected in the recent book, *The Seven Deadly Sins of Psychology* by Chris Chambers. Chapter 1, called "The Sin of Bias," starts right out with a bias of its own, the bias of physicalism. It asserts that Daryl Bem's experiments on "Feeling the Future" couldn't possibly have positive results because being aware of the future is impossible. The author uses the words "supernatural" and "ridiculous" to support his bias. He says "extraordinary claims" without apparently being aware of the bias that makes these claims appear "extraordinary." He calls the rebuttal of Bem's paper by Wagenmakers et al. a "statistical demolition" without explaining scientifically, logically, and mathematically exactly how that demolition was accomplished (Chambers, 2017, pages 1-4). Actually, I have demolished their use of Bayesian statistics in one sentence by saying, "In Bayesian analysis, they are able to plug their bias into the equation, and out the other end comes an answer that reflects their bias."

This book is published by Princeton University Press, thus using the name of Princeton University, the #1-ranked university in the country, to give it prestige. The smear-tactics used, combined with the high prestige of the name, make this chapter a perfect example of dirty science.

These people really don't know — the author, his many supporters, or the publisher. And yet they are so certain in their ignorance that it is an embarrassment.

I know. I am certain because I have experienced precognition myself. I have already described my precognitive dreams, my dream of the "camera car," and my positive results with the Bem experiment. I also recall that for about a year after smoking marijuana, I always knew when the traffic light was going to turn green. Only one such experience is required to know that such a thing as precognition is possible. Billions of people in the world have had similar experiences.

There is no question of the reality of precognition. These academic people are dead wrong. But they don't know that. Some self-knowledge is required, for the academic community even to recognize the in-group thinking, much less to break out of it. Self-knowledge suggests the teaching of Socrates, "Know thyself," and Freud's discovery of how to do that, namely psychotherapy. But it is risky to suggest psychotherapy to anyone, because people's natural defenses reject the idea unless they are already considering it themselves. And any suggestion of psychotherapy to academic people is even more likely to be rejected, because prejudice against psychotherapy for themselves is one of the in-group attitudes that is forced upon them, under threat of extreme ridicule, as I illustrated with physicist Wolfgang Pauli and his fear of the "hellish laughter."

Self-knowledge is not an academic subject. No degrees or credentials are issued for self-knowledge. The New School in New York briefly offered psychotherapy for credit in the 1960s, but that was all. The extreme prejudice keeps academic people from even thinking about psychotherapy for themselves, as if that puts them in the same category as "crazy people."

But as people in the field of psychotherapy have discovered over the years, the methods of Freud and Jung, even as modified and improved, don't work so well with "crazy people." They work much better with the "Type One Clients," as defined by Kenneth Howard and David Orlinsky in 1972 (Lazerson, 1975, page 593).

Any person in the normal range can benefit from psychotherapy, in all the ways I listed in Chapter 11, and those who reach psychological maturity will have a competitive advantage, both in their legitimate activities and in dealing with the politics of knowledge. I don't think that psychologists would have let themselves be steered away from the study of the mind if they had been psychologically mature. It may take something like mass psychotherapy to undo what I see as something like mass hypnosis in the academic world. But let's try first for some simple self-knowledge.

Self-knowledge requires some outside agency that is able to reflect the self. It can start with the same kind of insights as I had, trying to understand "The Waste Land" and recognizing that it was talking about me, my life, my world, and my social group. I am hoping that *Dirty Science* might do the same for scientific and academic people.

And then, for self-knowledge to grow, it requires constant feedback, as I had from my psychiatrist and my dreams. We, the public, need to send gentle reminders constantly from the outside, to help the members of the scientific and academic communities become aware of the social pressures limiting freedom of inquiry.

It should not be hard to recognize the dirty science, most of the time, by asking the three simple questions I listed in Chapter 1:

First, is this an illegitimate argument? Is the so-called "scientist" using smear words such as "fruitcake," "crackpot," "believer," "delusional," "silly," "hilarious," "myth," "stupid," and so on? This should be the easiest way to spot that it isn't science. Any kind of ridicule is not only not science, but is actually anti-science, in that it creates a bias, and any bias decreases the accuracy of science.

Second, is the so-called "scientific" opinion based on replicated scientific studies? Science always must be based on observations. We need to require that scientists cite scientific studies to back up all their assertions. If they can produce one such study, that makes it possible truth. If the study has been replicated many times, it is probable scientific truth. If they can't cite any scientific studies to back up their assertions, then those assertions are not scientific, and are no more likely to be true than something you might have heard from a guy in a bar. For example, their opinions that there are no such things as ghosts or angels, or that psychic abilities are impossible, have absolutely no scientific backing at all.

Third, is this person qualified in this particular field? The scientist's credentials should be published somewhere. Find out the scientist's credentials. Respect what he/she says within that field of expertise. But outside of that field, treat the scientist as knowing no more than you do. In particular, if a scientist is saying something derogatory about a field of study in which he/she is not qualified, that is just plain prejudice.

Predictably, if you have questioned their authority in these last two ways, the corrupt scientists may come at you with the authoritarian methods of status-snobbery, as in "Who are you to question?" You need to recognize that this is not science. They are trying to capitalize on the psychological conditioning we all have had in relation to authority figures. But a person's scientific credentials are not evidence that the person's opinion is scientific. And a person's scientific credentials in one field do not make that person an expert in all fields.

Or they may come at you with ridicule. You need to recognize that this also is not science, but a method of social manipulation and domination totally unrelated to science or the acquisition of accurate knowledge. You need to be able to stare down the ridicule. Or you might answer it as I plan to do, and say, "Thank you. That is a wonderful example of dirty science."

The ridicule is never necessary. It is an ego-prop. It is an ego-assertion: "We are so superior." If one is truly superior, one doesn't need the assertion.

Think of a first-grade teacher, whose knowledge is vastly superior to that of her students. She doesn't ridicule them, but patiently begins to teach them, "2 + 2 = 4," and "See Jane run."

So if a scientist is ridiculing you, think of it as the scientist's psychological ego-defenses. And if you think about it long enough, you will probably discover that this ridicule is probably to cover his/her ignorance in a field which he/she doesn't know anything about. The assertion "It doesn't exist" implies that the person knows nothing about the subject.

With these few rules, I think that an intelligent public will have enough weapons to successfully storm the Bastille, or at least make cracks in the walls.

As I already mentioned in Chapter 3, we can all, in our various ways, help the academic community to receive the self-awareness it needs.

The most obvious courses of action are those for people with money or who are responsible for disbursing large sums of money. There is no reason at all that you have to be manipulated by the ridicule, especially if you understand that it is not scientific but is actually anti-scientific. You have the power to withhold donations from institutions with extreme prejudices. And you might consider giving money specifically earmarked for parapsychology or spiritual studies, as was done at the University of Virginia to support reincarnation research.

I don't know much about accrediting committees, except that an accrediting committee once voted down the study of parapsychology at an unranked university because the subject was too "controversial." I am hoping that accrediting committees will give more weight to legitimate scientific evidence than to social pressures.

If you are a prospective student and interested in precognition, telepathy, clairvoyance, remote viewing, psychokinesis, energy medicine, spirit entities, the power of prayer, reincarnation, levitation, or intelligent design, you can politely inquire of colleges and universities whether they give courses in these subjects and what departments would be responsible for such courses. You can have adults such as your parents, a teacher, a yoga instructor, a minister, or Uncle Bobby back you up and stand up to them with authority if they respond with ridicule or condescension.

If you are an adult sending a child to college, for $60,000 a year and more, you can ask them what you are getting for your money, and whether your child's education will be limited to only the physical, or will it include also the mental, the emotional, the psychic, and the spiritual.

Anybody can send a letter to the president of a college or university requesting that they add non-physical subjects to their curriculum. They are responsive to their mail. In fact, one president advertised that he responded to every letter. He didn't.

Lawyers might be interested in dirty science. They have certainly managed to extract money from the medical profession for malpractice. Much of dirty science is misrepresentation and defamation. I think a person should be able to claim enough monetary damage to cover the cost of a lawyer to clear his/her good name. I am hoping that members of the legal profession will give this some thought.

Most important, sincere scientists and scholars might speak out, if their careers aren't jeopardized by doing so. Actually, a majority of academic people accepted the reality of psychic phenomena a long time ago (Bem & Honorton, 1994). The people doing the dirty science are actually a minority, but like the religious right, they exert an influence way out of proportion to their numbers. And, one more time, let us remember that scientists voted 5 to 1 in favor of parapsychology in 1969, after an enthusiastic recommendation by Margaret Mead. A respected scientist might have the same impact today.

To start with, every student of critical thinking should read this book.

I invite all of you to share your experiences with dirty science on the blog at dirtyscience.net, particularly any of you who may have been fired from a major university or shunned because of your interest in the psychic or the spiritual. We need more evidence from the inside that there is this kind of persecution going on. If enough of you have such evidence, I am hoping that you will co-author with me another book, describing your experiences.

Just the cultural awareness alone should be a major factor in eliminating dirty science from our academic environment. When the cracks in the defensive structure are revealed, it may crumble by itself.

I see an opening in the Humanities. The area of the Humanities doesn't have to be dominated by the views or prejudices of the scientific establishment. Right now philosophy and religious studies are included under the umbrella of the Humanities. It would not be a stretch to include the study of the spiritual, independent of religion, or psychotherapy for credit, as a philosophy course, or even the scientific exploration of the psychic and the spiritual, since these things are more related to philosophy than to the physical sciences. Of the three divisions of academic study — the Natural Sciences, the Social Sciences, and the Humanities — it seems that the Humanities should be the least dominated by the in-group views of physical scientists, because its name doesn't include the word "sciences."

Another possibility is that the unranked alternative universities might gain status, as students flock to them to learn about the spiritual and the psychic.

These are only possibilities. What will actually happen is up to you.

References and Bibliography

Alcock, J. E. (1995). The belief engine. *Skeptical Inquirer, 19*, 14-18.

Alcock, J. E. (2011). Back from the future: Parapsychology and the Bem affair. *Skeptical Inquirer*, Vol. 35 No. 2, 31-39.

Alloy, L. B., Jacobson, N. S., and Acocella, J. (1999). *Abnormal Psychology: Current Perspectives*. New York: McGraw-Hill.

Alvarado, C. S. (2014). Recent surveys of psychic experiences: II. Retrieved from: https://carlossalvarado.wordpress.com/2014/07/28/recent-surveys-of-psychic-experiences-ii/

Beloff, J. (1980). Seven evidential experiments. *Zetetic Scholar, 6*, 91-94.

Bem, D. J. (2011). Feeling the future: Experimental evidence for anomalous retroactive influences on cognition and affect. *Journal of Personality and Social Psychology, 100*, 407-425.

Bem, D. J., and Honorton, C. (1994). "Does psi exist? Replicable evidence for an anomalous process of information transfer." *Psychological Bulletin, 115*, 4-8.

Bem, D. J., Utts, J., and Johnson, W. O. (2011). Reply: Must psychologists change the way they analyze their data? *Journal of Personality and Social Psychology, 101*, 716-719.

Berelson, B., and Steiner, G. A. (1964). *Human Behavior: An Inventory of Scientific Findings*. New York: Harcourt, Brace & World.

Bernstein, D. A., Clarke-Stewart, A., Roy, E. J., and Wickens, C. D. (1997). *Psychology*, Fourth Edition. Boston: Houghton

Mifflin Company.

Bogart, G. (2004). Therapeutic dreamwork: A case study with mythic dimensions, Part 1. *Dream Network*, Vol 23 No 1: 38-43.

Boring, E. G. (1950). *A History of Experimental Psychology*. New York: Appleton-Century-Crofts.

Boring, E. G. (1953). A history of introspection. *Psychological Bulletin, 50*, 169-186.

Bowler, P. J. (1989). *Evolution: The History of an Idea*. Berkeley: University of California Press.

Bro, H. H. (1968). *Edgar Cayce on Dreams*. Edited by Hugh Lynn Cayce. New York: Paperback Library.

Broughton, R. S. (1991). *Parapsychology: The Controversial Science*. New York: Ballantine Books.

Buckley, W. F., Jr. (1951). *God and Man at Yale: The Superstitions of "Academic Freedom."* Chicago: Henry Regnery.

Burk, L. (2012). *Let Magic Happen: Adventures in Healing with a Holistic Radiologist*. Durham, NC: Healing Imager Press.

Cairns-Smith, A. G. (1985). *Seven Clues to the Origin of Life: A Scientific Detective Story*. Cambridge, England: Cambridge University Press.

Cardeña, E. (2011). Guest Editorial: On wolverines and epistemological totalitarianism. *Journal of Scientific Exploration, 25*, 539-551.

Cardeña, E. (2013). Book Review: *Science and Psychic Phenomena: The Fall of the House of Skeptics*, by Chris Carter. *Journal of Scientific Exploration, 27*, 520-524.

Cardeña, E. (2015). The unbearable fear of psi: On scientific censorship in the 21st century. *Journal of Scientific Exploration, 29*, 601-620.

Cayce, E. E., and Cayce, H. L (1971). *The Outer Limits of Edgar Cayce's Power*. New York: Harper & Row.

Cayce, H. L. (1964). *Venture Inward*. New York: Paperback Library.

Cerminara, G. (1950). *Many Mansions*. New York: William Morrow.

Chambers, C. (2017). *The Seven Deadly Sins of Psychology: A*

Manifesto for Reforming the Culture of Scientific Practice. Princeton, NJ: Princeton University Press.

Cohen, S. (1967 [1966]). *The Beyond Within: The LSD Story.* New York: Atheneum.

Committee for the Scientific Investigation of Claims of the Paranormal (CSICOP). (1996). *The Outer Edge: Classic Investigations of the Paranormal.* Edited by Joe Nickell, Barry Karr and Tom Genoni. Amherst, NY: CSICOP.

Coyne, J. A. (2009). *Why Evolution Is True.* New York: Penguin.

Crews, F. (1975). *Out of My System: Psychoanalysis, Ideology, and Critical Method.* New York: Oxford University Press.

Crews, F., et al. (1995). *The Memory Wars: Freud's Legacy in Dispute.* New York: The New York Review of Books.

Dawkins, R. (1987). *The Blind Watchmaker: Why the evidence of evolution reveals a universe without design.* New York: W. W. Norton.

Dawkins, R. (1992). Fossil fool, a review of *The Facts of Life: Shattering the Myth of Darwinism,* by Richard Milton. *New Statesman & Society,* Vol. 5, No. 217, Aug. 28, 1992, pages 33-34.

Deresiewicz, W. (2014). *Excellent Sheep: The Miseducation of the American Elite and the Way to a Meaningful Life.* New York: Simon & Schuster.

Descartes, R. (1977 [1637]). *Discourse on Method.* Indianapolis, IN: Bobbs-Merrill.

Druckman, D., and Swets, J. A., Editors (1988). *Enhancing Human Performance: Issues, Theories, and Techniques.* Washington, DC: National Academy Press.

Eldredge, N. (1985). *Time Frames: The Rethinking of Darwinian Evolution and the Theory of Punctuated Equilibria.* New York: Simon and Schuster.

Eliot, T. S. (1962 [1922]). The waste land. In *The Waste Land and Other Poems.* New York: Harcourt Brace Jovanovich.

Elkins, D. N. (1998). *Beyond Religion: A Personal Program For Building A Spiritual Life Outside The Walls Of Traditional Religion.*

Wheaton, IL: Theosophical Publishing House.

Feather, S. R., and Schmicker, M. (2005). *The Gift: ESP, the Extraordinary Experiences of Ordinary People*. New York: St. Martin's Press.

Feyerabend, P. (2002 [1987]). *Farewell to Reason*. London: Verso.

Feyerabend, P. (2010 [1975]). *Against Method*, Fourth Edition. London: Verso.

Franzoi, S. L. (1996). *Social Psychology*. Dubuque, Iowa: Brown & Benchmark.

Gallup (2016). Religion. Retrieved from: www.gallup.com/poll/1690/ religion.aspx

Garfield, S. L., and Bergin, A. E. (1986). *Handbook of Psychotherapy and Behavior Change*. New York: John Wiley & Sons.

Gebelein, B. (1985). *Re-Educating Myself: An Introduction to a New Civilization*. Provincetown, MA: Omdega Press.

Gebelein, B. (2007). *The Mental Environment: (Mostly about Mind Pollution)*. Provincetown, MA: Omdega Press.

Gebelein, R. S. (2013). Guest Editorial: Physicalism. *Journal of Parapsychology, 77*, 159-164.

Gould, S. J. (1999). *Rocks of Ages: Science and Religion in the Fullness of Life*. New York: Ballantine.

Hanegraaff, H. (2003). *Fatal Flaws*. Nashville: Thomas Nelson.

Hansel, C. E. M. (1980). *ESP: A Scientific Evaluation*. New York: Scribner.

Hess, D., and Layne, L., Editors (1992). Disciplining heterodoxy, circumventing discipline: Parapsychology, anthropologically. In *Knowledge and Society: The Anthropology of Science and Technology*, Volume 9, pages 223-252. Greenwich, CT: JAI Press.

Homans, G. C. (1950). *The Human Group*. New York: Harcourt, Brace & World.

Homans, G. C. (1961). *Social Behavior: Its Elementary Forms*. New York: Harcourt, Brace, & World.

Horn, S. (2009). *Unbelievable: Investigations into Ghosts, Poltergeists, Telepathy, and Other Unseen Phenomena, from the Duke Parapsychology Laboratory*. New York: HarperCollins.

Howard, K. I., and Orlinsky, D. E. (1972). "Psychotherapeutic

Processes," in Paul Mussen and Mark Rosenweig (eds.), *Annual Review of Psychology: XXIII*. Palo Alto, CA: Annual Reviews.

Huxley, T. H. (1871). *Lay Sermons, Addresses, and Reviews*. New York: Appleton.

Illustra Media (2009). *Darwin's Dilemma* [DVD]. Orange, CA: Carmel Entertainment Group.

James, King (1955 [1611]). *The Holy Bible*, Authorized King James Version. New York: Oxford University Press.

Janis, I. L. (1982). *Groupthink: Psychological Studies of Policy Decisions and Fiascoes*. Boston: Wadsworth.

Johns, R. J. (1987). How to swim with sharks: A primer by Voltaire Costeau. *Perspectives in Biology and Medicine, 30*, 486-489. Retrieved from: www.med.upenn.edu/shorterlab/Papers/Member%20Papers/sharks.pdf

Johnson, P. E. (2010 [1991]). *Darwin on Trial*. Downers Grove, IL: InterVarsity Press.

Jung, C. G. (1959). *The Basic Writings of C.G. Jung*. Edited by Violet S. de Laszlo. New York: The Modern Library.

Jung, C. G. (1983). *The Essential Jung*. Princeton, NJ: Princeton University Press.

Kant, I. (1997 [1781]). *Critique of Pure Reason*. Edited and translated by Paul Guyer and Allen W. Wood. Cambridge, England: Cambridge University Press.

Karagulla, S. (1967). *Breakthrough to Creativity: Your Higher Sense Perception*. Los Angeles: DeVorss.

Kastrup, B. (2014). *Why Materialism Is Baloney: How true skeptics know there is no death and fathom answers to life, the universe and everything*. Winchester, UK: iff.

Kieninger, R. (1971). *Observations*. Stelle, IL: The Stelle Group.

Kieninger, R. (1974a). *Observations II*. Stelle, IL: The Stelle Group.

Kieninger, R. (1974b). *Observations III*. Stelle, IL: The Stelle Group.

Kieninger, R. (1978). Ten Qualities of Mind, in *On Becoming an Initiate* [lecture series]. Stelle, IL: The Stelle Group.

Kieninger, R. (1979). *Observations IV*. Stelle, IL: The Stelle Group.

Kieninger, R. (1986). *Spiritual Seekers' Guidebook: And Hidden Threats to Mental & Spiritual Freedom*. Quinlan, TX: The Stelle Group.

Kueshana, E. (1963). *The Ultimate Frontier*. Stelle, IL: The Stelle Group.

Lamsa, G. M. (1957 [1933]). *The Holy Bible*, translated by George M. Lamsa. Philadelphia: A. J. Holman.

Larson, E. J., and Witham, L. (1998). Leading scientists still reject God. *Nature*, Vol. 394, No. 6691, page 313. Retrieved from: www.nature.com/nature/journal/v394/n6691/full/394313a0. html

Lazarsfeld, P. F., and Thielens, W., Jr. (1958). *The Academic Mind: Social Scientists in a Time of Crisis*. Glencoe, IL: The Free Press.

Lazerson, A., Editor (1975). *Psychology Today: An Introduction*, Third Edition. New York: CRM / Random House.

Lett, J. (1992). The persistent popularity of the paranormal. *Skeptical Inquirer, 16*, 381-388.

Locke, J. (1975 [1690]). *An Essay Concerning Human Understanding*. Edited by Peter H. Nidditch. Oxford: Oxford University Press.

Luborsky, L., Chandler, M., Auerbach, A.H., Cohen, J., and Bachrach, H.M. (1971). Factors influencing the outcome of psychotherapy: A review of quantitative research. *Psychological Bulletin, 75*, 145-185.

Maslow, A. H. (1966). *The Psychology of Science: A Reconnaissance*. Chicago: Henry Regnery.

Masters, R. A. (2013). Spiritual bypassing: Avoidance in holy drag. Retrieved from: robertmasters.com/writings/spiritual-bypassing

Mathieu, I. (2011). Beware of spiritual bypass. Retrieved from: www.psychologytoday.com/blog/emotional-sobriety/ 201110/beware-spiritual-bypass

Mauskopf, S. H., and McVaugh, M. R. (1980). *The Elusive Science: Origins of Experimental Psychical Research*. Baltimore: The Johns Hopkins University Press.

McClenon, J. (1984). *Deviant Science: The Case of Parapsychology.* Philadelphia: University of Pennsylvania Press.

McClenon, J., Roig, M., Smith, M. D., and Ferrier, G. (2003). The coverage of parapsychology in introductory psychology textbooks: 1990-2002. *Journal of Parapsychology, 67,* 167-179.

Merriam-Webster (1996). *Merriam-Webster's Collegiate Dictionary.* Springfield, MA: Merriam-Webster, Incorporated.

Merriam-Webster (2017). "Introspection". Retrieved from: http://www.merriam-webster.com/dictionary/introspection

Miller, L. (2015). *The Spiritual Child: The New Science on Parenting for Health and Lifelong Thriving.* New York: St. Martin's.

Milton, R. (1996 [1994]). *Alternative Science: Challenging the Myths of the Scientific Establishment.* Rochester, VT: Park Street Press.

Milton, R. (1997 [1992]). *Shattering the Myths of Darwinism.* Rochester, VT: Park Street Press.

Mins, H. F., Jr. (1934). *Materialism: The Scientific Bias.* New York: Columbia University.

Oxford University Press (1964 [1933]). *The Shorter Oxford English Dictionary on Historical Principles.* Prepared by William Little. Revised and edited by C. T. Onions. Oxford, England: Clarendon Press.

Palmer, J. (1978). Extrasensory perception: Research findings. In *Advances in Parapsychological Research, 2: Extrasensory Perception.* Edited by Stanley Krippner. New York: Plenum.

Palmer, J. (1987). Dulling Occam's Razor: The role of coherence in assessing scientific knowledge claims, *European Journal of Parapsychology, 7,* 73-82.

Palmer, J. (2011). EDITORIAL: On Bem and Bayes. *Journal of Parapsychology, 75,* 179-184.

Palmer, J., Honorton, C., and Utts, J. (1988). *Reply to the National Research Council Study on Parapsychology.* Columbus, OH: Parapsychological Association.

Parapsychological Association (2015). Where can I get a degree in parapsychology? Retrieved from: http://parapsych.org/articles/36 /47/where_can_i_get_a_degree_in.aspx

Park, R. L. (2000). *Voodoo Science: The Road from Foolishness to Fraud.* Oxford, England: Oxford University Press.

Peck, M. S. (1978). *The Road Less Traveled: A New Psychology of Love, Traditional Values and Spiritual Growth.* New York: Simon & Schuster.

Pew Research Center (2009). Scientists and belief. Retrieved from: www.pewforum.org/2009/11/05/scientists-and-belief/

Putnam, C. (2014). Paranormal is the new normal. Retrieved from: www.supernaturalworldview.com/2014/05/14/paranormal-is-the-new-normal/

Radin, D. (2013). *Supernormal: Science, Yoga, and the Evidence for Extraordinary Psychic Abilities.* New York: Random House.

Randi, J. (1982). *FLIM-FLAM!: Psychics, ESP, Unicorns and Other Delusions.* Buffalo, NY: Prometheus Books.

Rao, K. R., and Palmer, J. (1987). The anomaly called psi: Recent research and criterion. *Brain and Behavior Sciences, 10,* 539-643.

Rhine Research Center (2001). *Parapsychology and the Rhine Research Center.* Durham, NC: The Parapsychology Press.

Roe, C. (2016). What Are Psychology Students Told About the Current State of Parapsychology? *Mindfield: The Bulletin of the Parapsychological Association, 7,* 86-91.

Russell, B. (1923). *A Free Man's Worship.* Portland, ME: Thomas Bird Mosher.

Schmicker, M. (2002). *Best Evidence: An Investigative Reporter's Three Year Quest to Uncover the Best Scientific Evidence for ESP, Psychokinesis, Mental Healing, Ghosts and Poltergeists, Dowsing, Mediums, Near Death Experiences, Reincarnation, and Other Impossible Phenomena That Refuse to Disappear.* Lincoln, NE: iUniverse.

Schmidt, H. (1969). Precognition of a quantum process. *Journal of Parapsychology, 33,* 99-108.

Schultz, D. P. (1969). *A History of Modern Psychology.* New York: Academic Press.

Schwitzgebel, E. (2016, Winter). "Introspection", The Stanford Encyclopedia of Philosophy. Edited by Edward N. Zalta. Retrieved from: https://plato.stanford.edu/archives/

win2016/entries/ introspection/

Sheldrake, R. (2012). *The Science Delusion: Freeing the Spirit of Enquiry*. London: Hodder & Stoughton.

Simpson, G. G. (1983). *Fossils and the History of Life*. New York: Scientific American Library.

skepdic.com (2016). Edgar Cayce (1877-1945). Retrieved from: skepdic.com/cayce.html

Smith, M. L., Glass, G. V., and Miller, T. I. (1980). *The Benefits of Psychotherapy*. Baltimore: Johns Hopkins University Press.

Stevenson, I. (1966). *Twenty Cases Suggestive of Reincarnation*. New York: American Society for Psychical Research.

Sugrue, T. (1967 [1942]). *There Is a River: The Story of Edgar Cayce*. New York: Dell.

Szasz, T. (1961). *The Myth of Mental Illness*. New York: Harper & Row.

Szasz, T. (1974). *The Myth of Mental Illness: Foundations of a Theory of Personal Conduct*. Revised edition. New York: Harper & Row.

Szasz, T. (1988 [1978]). *The Myth of Psychotherapy: Mental Healing as Religion, Rhetoric, and Repression*. Syracuse, NY: Syracuse University Press.

Szasz, T. (1990 [1976]). *Anti-Freud: Karl Kraus's Criticism of Psychoanalysis and Psychiatry*. Syracuse, NY: Syracuse University Press.

Tart, C. T. (2009). *The End of Materialism: How Evidence of the Paranormal Is Bringing Science and Spirit Together*. Oakland, CA: New Harbinger.

Thoreau, H. D. (1937 [1854]). *Walden*. In *Walden and Other Writings of Henry David Thoreau*. Edited by Brooks Atkinson. New York: Modern Library.

Thornton, E. M. (1986 [1983]). *The Freudian Fallacy: Freud and Cocaine*. London: Paladin Books.

Toynbee, A. J. (1957 [1946]). *A Study of History*. Edited by D. C. Somervell. New York: Dell.

Tsakiris, A. (2014). *Why Science Is Wrong ... About Almost Everything*. San Antonio: Anomalist Books.

Utts, J. (1996). An assessment of the evidence for psychic functioning. *Journal of Scientific Exploration, 10,* 3-30.

van Erkelens, H. (2002). Wolfgang Pauli, the feminine and the perils of the modern world. *Harvest: Journal for Jungian Studies, 48,* 142-148.

Velikovsky, I. (1965 [1950]). *Worlds in Collision.* New York: Dell.

Wagenmakers, E. J., Wetzels, R., Boorsboom, D., and van der Haas, M. (2011). Why psychologists must change the way they analyze their data: The case of psi. *Journal of Personality and Social Psychology, 100,* 426-433.

Watson, J. B. (1913). Psychology as the behaviorist views it. *Psychological Review, 20,* 158-177.

Watson, J. B. (1919). *Psychology from the Standpoint of a Behaviorist.* Philadelphia: J.B. Lippincott.

Watson, J. B. (1925 [1924]). *Behaviorism.* New York: W.W. Norton.

Watts, A. (1966). *The Book; On the Taboo against Knowing Who You Are.* New York: Pantheon Books.

Welwood, J., and Fossella, T. (2011). Human nature, Buddha nature: On spiritual bypassing, relationship, and the dharma; An interview with John Welwood by Tina Fossella. Retrieved from:
www.johnwelwood.com/articles/TRIC_interview_uncut.doc

Wikipedia (2016, January 15). Parapsychology. Retrieved from:
https://en.wikipedia.org/wiki/Parapsychology

Wilson, E. O. (1998). *Consilience: The Unity of Knowledge.* New York: Alfred A. Knopf.

Winn, D. (1983). *The Manipulated Mind: Brainwashing, Conditioning and Indoctrination.* London: The Octagon Press.

Wolman, B. B., Editor (1977). *Handbook of Parapsychology.* New York: Van Nostrand Reinhold.

Zingrone, N. L. (2002). 2001 PRESIDENTIAL ADDRESS: Controversy and the problems of parapsychology. *Journal of Parapsychology, 66,* 3-30.

Index

def = defined
many = many references, not all
listed
Quotes mean "so-called."